DEDICATION

To……

My Beloved Iraq

ACKNOWLEDGMENT

Thanks to Allah for providing us with power and patience to achieve this work.

We would like to thank the Ministry of Higher Education and Scientific Research in Republic of Iraq, for giving us the chance to write this book. Moreover, it is our great pleasure to thank the Colleges of Engineering in Al-Nahrain University, Al-Muthanna University and Middle Technical University that supported us during years of authorship of this book.

We would like to extend our deepest respect and appreciation to Assistance Professor Dr. Ibrahim A. Mohammed for his encouragement, patience, meaningful, and unlimited cooperation during writing this book.

We would like to express our gratitude to the many people who provided support, talked things over, read, wrote, offered comments, allowed us to quote their remarks and assisted in the editing, proofreading and design.

Grateful thanks and appreciation to our families for their continuous support to perform this book. Last and not least: We beg forgiveness of all those who have been with us over the course of the years and whose names accidently failed to mention.

Authors

LIST OF ABBREVIATIONS

ASTM	American Society for Testing and Materials
ASSHTO	American Association of Highway Transportation Officials
APWA	American Public Works Association
CP	Composite Pavement
CBR	California Bearing Ratio
DSS	Decision Support System
FAA	Federal Aviation Administration
FP	Flexible Pavement
IRI	International Roughness Index
JRC	Jointed Reinforced Concrete
LoS	Level of Service
M&R	Maintenance and Rehabilitation
MCI	Maintenance Condition Index
PD	PAVEMENT DETERIORATION
PM	Project Management
PCI	Pavement Condition Index
PSR	Pavement Serviceability Rating
PSI	Present Serviceability Index
PQI	Pavement Quality Index
PMMS	Pavement Maintenance Management System
RCC	Roller-Compacted Concrete
RP	Rigid Pavement
SN	Skid Number

LIST OF CONTENTS

Chapter Two

HIGHWAY PROJECTS

Chapter Three

PAVEMENT DETERIORATION

Chapter Four

HIGHY WAY MAINTENANCE OPERATIONS

Chapter Five

PAVEMENT MAINTENANCE MANAGEMENT

LIST OF FIGURES

LIST OF TABLES

PREFACE

This is the first edition of a book on highway maintenance management which is designed specifically for the needs of division undergraduate students in civil engineering in particular. This book is designed to introduce highway maintenance management concepts to any construction project professional who is considering the implementation of pavement project for the first time. Whether you're a seasoned project manager, a highway project maintenance professional, or an executive this book will give you practical advice on making this journey successful.

No assumptions have been made about previous experience with either highway project management or pavement maintenance, thus providing those who have only recently found themselves operating in an highway environment the ability to become valued contributors to the project management function.

This book was written by academic's and practitioners who have lead the implementation of highway management processes and tools at several major corporations. The contents of this book have been presented in an interesting and enjoyable way, enhanced by real pictures of highway projects and pavement maintenance. This book contains six chapters, the first chapter entitled **MAINTENANCE MANAGEMENT**: It was to clarify the concept and importance of maintenance and management professionally and smoothly, While the title of the second chapter is the **HIGHWAY PROJECTS**, and provided a detailed explanation of the management and implementation of highways, while reviewing the types and importance in the construction sector. The third chapter, entitled PAVEMENT DETERIORATION: The researchers reviewed the types of DETERIORATION in the riged and asphalt pavement, and explained the methods of treatment and maintenance necessary for each type. While the fourth chapter was entitled HIGHY WAY MAINTENANCE OPERATIONS: It reviewed the methods of maintenance and importance in highway project, the fifth chapter entitled: PAVEMENT MAINTENANCE MANAGEMENT SYSTEM: This chapter reviewed the most important global strategies in the management of pavement maintenance.

The purpose of this book is to collect the diasporas of these tools and methods scattered in the stomachs of original of books and periodicals to be organized in a single travel without elaborating or prolonging, commenting on what needs comment, and our adoption after God almighty on the many of references we have proven it in the list of references in order to the interest of scientific honesty on the one hand and to benefit from it, and to return to it without exerting effort and wasting time. Finally, we praise the God for the great bounty that gave us the ability to accomplish this book, so it may be beneficial to the engineers and help our students in their lesson.

CHAPTER ONE

1 MANAGEMENT OF MAINTENANCE

1.1 Introduction

Maintenance of any structure involves administrative and scientific techniques, which includes supervising the structure after completion of implementation and throughout its life in order to keep the structure systems and components such as architectural, structural and mechanical operable or to rehabilitate structures so that they become usable within the proposed function.

1.2 Management of Maintenance

The management of maintenance can be understood as standardization of maintenance activities and resources necessary to achieve and maintain effective control of the structure to ensure the best use of information technology at reasonable cost and acceptable levels.

All structures are facing deterioration with the passage of time, so is a necessity to make a comprehensive maintenance for these structures to protect them from all types of deterioration and damage that make them unsuitable for use. These also can be associated with threats to health and safety of the occupants. Many factors may affect the maintenance requirements including:

1) Age of structures.
2) Usage of structures.
3) Environmental and structural conditions.
4) Finishing.
5) Materials used.
6) Level of quality before and after construction.

7) Level of maintenance required for the structure and the provision of support and funding.

In addition, high-cost structures require preservation and maintenance of all the influences that might lead to a reduction in its designated life.

Hence, there is an urgent need for clear information about the comprehensive maintenance of various cases.

Structures are usually maintained by conducting various types of proper maintenance for all components of structures. Problems may occur in case of ignoring or postponing the required maintenance in order to reduce expenses.

The fundamental objective of all types of maintenance is to reduce projected spending on structures by knowing and identifying required maintenance for these structures. Therefore, the creation of an integrated management system for the maintenance work is very essential .

Advantages of maintenance works performed on structures may include:

1) Better performance of structures.
2) Extension of the service life of structures.
3) Minimizing operating costs.
4) Better book value of maintained structures.

1.2.1 Importance of Maintenance Management

The management of maintenance is very important for the following reasons:

1) Management of cost of maintenance activities represents 30-50% from the total cost of production.
2) Optimum use of the structure and an increase in its service life.
3) Optimal use of resources (fuel, electricity, etc.) and rationalizing them.

The high level of technology in most of the industries, nowadays, requires advanced engineering information and high skills for implementing maintenance works.

1.2.2 Challenges of Maintenance Management

The maintenance management faces several challenges and problems depending on the type of technical mechanism employed and its degree of complexity.

This requires an administration to develop special standards and bases in terms of the followings:

1) Periods of examination and inspection.

2) Dates of major maintenance works.

3) The size of the treatment maintenance needed to restore a servic to its natural state.

4) The size of the required maintenance teams.

5) The quantity of services provided by workshops.

6) Plans and policies used in performing of maintenance operations

1.3 Definitions of Maintenance

Some of the most known definitions of maintenance are liste below:

1) "A group of technical and administrative actions performed t keep the assets away from any failure that might reduce th efficiency and productivity ".

2) "All actions carried out to maintain and protect the structure fron possible damage so the structure performs its function as long a possible and hence reducing costs caused by the negligence".

3) "All implemented actions that lead to protecting the building or rehabilitating its components or the construction materials used in order to maintain its function within the criteria ".

Based on these three definitions of maintenance, three outcomes can be reached:

a. Maintenance activities have significant influences on structures; the effectiveness and success of the maintenance largely depends on the adequacy and efficiency of the plans prepared.

b. The maintenance system affects other systems in the institution; also it is affected by those systems.

c. The objective of effective maintenance is to keep the efficiency of the equipment of the property

1.4 Objectives of Maintenance

Objectives of maintenance should be consistent with the service objectives of the structure and this is reflected through protecting of components and facilities and keeping them functional as much as possible. The following figure (1.1) illustrates the most important objectives of maintenance.

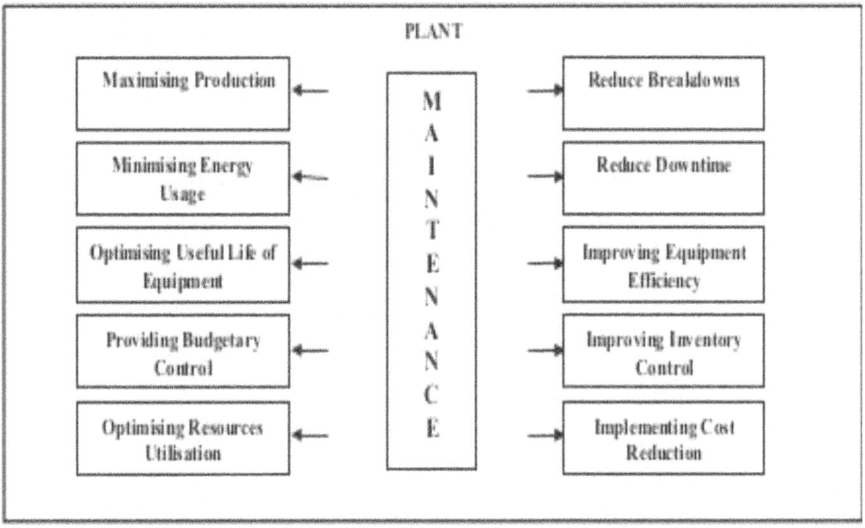

Figure (1.1): Objectives of Maintenance

1.4.1 Factors Identifying Objectives of Maintenance

Determining the objectives of maintenance is a quite complicated task because it requires cooperation among the operating, management and safety departments. Then, needs of maintenance should be thoroughly discussed with the maintenance administration. Figure (1.2) illustrates the factors that identify the objectives of maintenance.

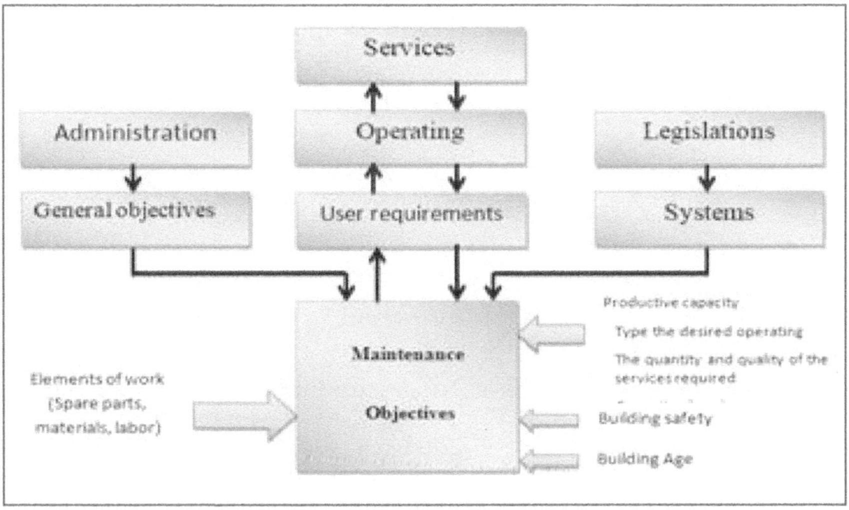

Figure (1.2): Factors Identifying Maintenance Objective

1.4.2 Achievements of Maintenance Objectives

The administration of the institution should be sure that the maintenance unit effectively achieves their specific targets. Therefore, effective, efficient and good regulatory systems should be set up. Such regulations are the followings:

1) Monitoring the discipline of workers in the maintenance department and their achievements.
2) Observing the logical work and its indications that may reflect to what level the staff works within the rules and instructions.
3) Monitoring losses caused by the breakdown of equipment and machines.
4) Collecting sufficient information regarding the situation, including the condition of maintenance and procedures of work.
5) Ensuring that is maintenance works are carried out according to right plans and specific timetable. These plans of action should be considered by comprehensiveness and coordination and collaboration between the various related units.

6) Providing a sufficient number of manpower, machinery and materials commensurate with the size and nature of the institution and the services provided.

1.5 Maintenance Activities

The size of maintenance activities varies from one work site to another and it is greatly affected by the size of the structure and the used techniques. The overall maintenance activities can be divided into major and minor activities.

1) Major activities include:
   a) Maintenance of buildings and plazas.
   b) Maintenance of machinery and equipment.
   c) Modifications or additions to the designs of buildings.
   d) Maintenance of water and sewer extensions.
   e) Examination and inspection activities.
2) Minor activities include:
   a) Occupational Safety.
   b) Administration of store backup materials.
   c) Waste disposal.

PROTECT
PRESERVE
KEEP
Maintenance
to Maintain, Keep, Preserve & Protect

1.6 Types of Maintenance

Figure (1.3) shows types of maintenance for any structure.

1) Planned maintenance: organizing the maintenance work in advance and controlling it according to prepared estimates and documented procedures. The maintenance work is carried out according to preset plans
2) Unplanned maintenance: Activities of maintenance occur without pre- planning.
3) Preventive maintenance: Activities of maintenance that are carried out according to plans and within the schedule for the purpose of decreasing the possibility of failure and enhancing the performance.
4) Corrective Maintenance: Activities of maintenance that take place after happening of the failure. This aims at restoring the situation back to normal conditions.
5) Scheduled maintenance: This maintenance is a preventive maintenance through which activities are carried out as per defined specific quantitative criteria such as mileage, periods of time and number of operating times.

6) Condition-based Maintenance: It is also a preventive maintenance through which activities are promoted according to outcomes of the routine tests and inspection.

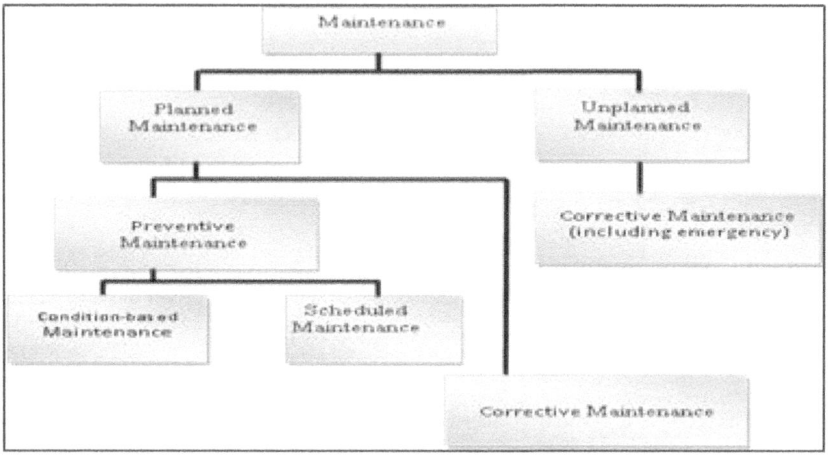

Figure (1.3): Types of Maintenance

1.7 Maintenance Priorities

Maintenance priorities include giving high priority to what would produce real risk of the structure or its contents and as follows:

1) The first priority: the dangers that might affect the lives of people, their environment and the health conditions related to them.
2) The second priority: the dangers that might affect the continuity of works within the facility.

1.8 Properties of Maintenance Type Selection

Selection the method of the required maintenance can be relied upon three basic properties:

1) Failure properties: This method involves analyzing the types of malfunctions, their frequency, and studying the possible causes in order to determine the optimum time of maintenance, the possibilities of performing it and the availability of its resources (labor and spare parts). This includes:
   a) The average time elapsed until the occurrence of malfunctions.
   b) The average age of the piece where the malfunction occurred.
   c) The frequency of finding out the malfunction causes.
   d) The study inspection and replacement times and the possibility of reform.

2) Safety properties: Determining the properties of safety related to maintenance operations to reach better safety. This is based on the following: -

   a) The effect of malfunctions on safety.

   b) The allowable limit in the safety system associated with the age of equipment or their parts.

3) Cost Properties: Determining the properties of the cost in order to reach minimum cost as much as possible without losing the level of maintenance. This usually requires realizing the repetitions of malfunctions, times of downtime and repairing and also the labor required. Generally, the cost can be defined based on two essential elements, namely:

   a) Cost of downtime: It includes the cost lost in productivity during repairs, preventive maintenance or waiting for repairs.

   b) Maintenance cost: It includes costs of performing maintenance work which covers fixed and variable costs of labor, parts and energy. Figure (1.4) shows the relationship between the cost and the level of maintenance.

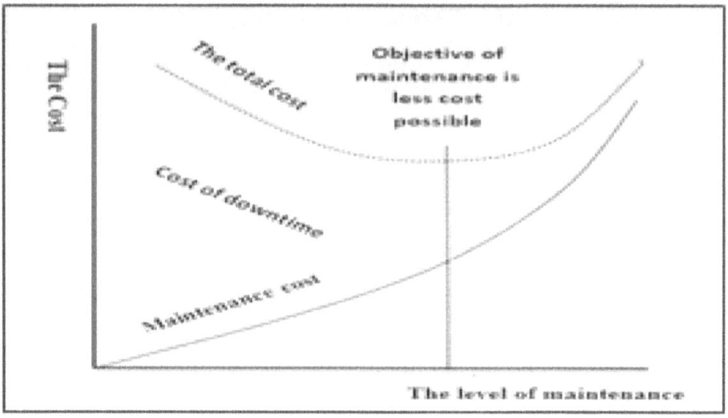

Figure (1.4): Level of Maintenance

1.9 Ways of Implementing Maintenance

There are four ways to carry out maintenance works and generally can be summarized as follows:

1) Maintenance by specialized units: Maintenance is carried out by a specialized unit in a specific maintenance work; for example, electricity unit, a medical equipment unit and etc. . The major advantages for this method are the ease of work, control and the opportunities for creativity, while the disadvantages may include the lack of utilization of available resources.

2) Maintenance by specialists: The maintenance staff consists of technicians from different disciplines. The advantages of this method are maximizing the use of available resources and the ease and quick coordination to carry out various maintenance. In contrast, the most key disadvantages are the low completion rate and the difficulty of supervision and control.

3) Maintenance by contractors: in this method, the maintenance is performed through contracts. Requirements and conditions of maintenance should be clarified in the contract. The typical disadvantages of this method are the high cost, the low level of control and the difficulty of getting the job done in line with the inspirations of the institution.

1.10 Maintenance Resources

There are two major resources required to carry out maintenance works:

1) Manpower: It concerns all issues regarding the management of personnel and the recruitment specifications and experiences of workers and their wages and incentives
2) Materials and spare parts: They include issues relating to the supply of tools and spare parts and also organizing and controlling of warehouses.

1.10.1 Human Resources

Human resources are the cornerstone for organizing maintenance works regardless the level of complexity of the maintenance work. If human resources are not able to run the system, there will be a real trouble. This could rationalize the importance of taking a number of decisive measures regarding the workers efficiency and available training programs required for improving the efficiency. The essential requirement in the maintenance job is honesty and truthfulness. This requirement could be achieved by taking two initiatives: the first is by running adequate motivating programs for employees while the second is by conducting appropriate training programs for each group of human resources, which include:

1) Skilled and non-skilled labors.
2) Technicians and Engineers.
3) Managers of work.
4) Trainees.

The training system usually includes; The safety, Administration of individuals, Maintenance according to the case, Transport vehicles of work, Planning for the task, Welding, Electronic-Business, Technical meetings and Others

Additionally, the precise and correct description of the maintenance job is very important to complete the maintenance work. .

By this way every worker should know all the conditions relating to the work mission. The number of people to be employed in the organization of maintenance varies from one place to another. One of the important things to be decided during the planning phase of maintenance is the size of the workforce and, of course, this has an impact on the training of human resources.

It can be concluded, based on the review of relevant literature, that the use of skilled labor leads to the following:

1) Rising in the operating efficiency of the units.
2) Reducing the damages during the maintenance process.
3) Decreasing the number of spare parts consumed.
4) Reducing the time required for maintenance and hence the completion date would be within the specified timetable.
5) Being fully prepared to cope with emergency conditions an critical situations.

In any case, it is a must that labors of all the required disciplines are available to implement the maintenance work adequately.

1.10.2 Materials and Spare Parts

It is realized that every part in a machine has a specific job to do during the machine useful life. This job is supposed to be done properly especially when running the machine under the circumstances and the conditions set by the machine factory.

There is no doubt that the provision of the necessary materials for a maintaining work (such as spare parts and lubricating oils, wires and cleaning materials etc..) has a direct impact on the success of the maintenance according to the institution's plans. Also, this aids in implementing the maintenance without delay.

Lack of spare parts may adversely affect the maintenance process; that is by increasing the duration in which the machine is out of work. This would reflect at the end the weakness of the institution in performing its function (Service or Industrial). In order to provide spare parts, it is important to follow the plan of purchase based on the following:

1) Identifying the types of spare parts.
2) Determining the sizes of the required spare parts.
3) Determining the demand size on spare parts.

CHAPTER TWO

2 HIGHWAY PROJECTS

2.1 Introduction

For as long as the human race has existed, transportation has played a significant role by facilitating commerce, trade, conquest, and social interaction,. The primary need for transportation involves personal travel in search of food or work, travel for the exchange of goods and commodities, exploration, personal fulfillment, and the improvement of societies or nations. The travel of people and goods, which is the basis of transportation, always has been considered to accomplish those objectives or tasks that require transfer from one location to another. The history of transportation illustrates that the way people move is affected by technology, cost, and demand. Transportation is a significant element in the daily life, also it represents approximately 18% of household expenditure and employing about 10% of the workforce.

2.2 Importance of Highways

Highways have a vital importance to the development of the country's economy. The construction of highway network with a high quality directly increases the nation's economic outputs by reducing journey times and costs, and making a region more attractive. The actual construction process have positive effect on stimulating the construction market

2.3 Selection of Highway Locations

Selecting the location of a proposed highway is a necessary as an initial step in design. The selection of a particular location is logically depending on topography of the region, environmental factors including noise, soil characteristics, air pollution, and economical factors. The required data for the decision process is usually obtained from different kinds of surveys, considering many factors. Engineering consultants and state agencies involved in selection of highway locations normally use computerized techniques to process the huge amounts of data that are handled in the decision process.

These techniques may include remote sensing, which uses aerial photographs for the preparation of maps, and computer graphics, which is a combination of the analysis of computer-generated data with a display on a computer monitor. The highway location should satisfy many requirements including; the easy flow of traffic at the design capacity considering meeting of design criteria and safety standards.

The highway location should also reduce disruption to historic and archeological sites and to other land-use activities. Therefore, the studies of environmental impact are needed in most cases as an early stage prior to approving of the highway location The process of selecting the highway location involves four phases:

1) The office study of available information.
2) Reconnaissance survey.
3) Survey of the proposed preliminary location.
4) Survey of the final selected location.

2.3.1 Office Study of Available Information

The first step in any highway location project is the examination of all existing data of the proposed area in which the highway will be constructed. This phase is performed in the office before any field or photogrammetric investigation. All available data are gathered and examined. The sources of data could be engineering reports, maps, aerial photographs, and charts These sources can be found in governmental offices such as departments of transportation, agriculture, geology, hydrology, and mining. The amount and type of collected data normally depend on the type of highway, but in general, data should be collected based on the following criteria:

1) Engineering data: This may include topography, geology, climat and traffic volumes
2) Social and demographic data: This may include land use an zoning patterns.
3) Environmental data which normally includes types of wildlife, location of recreational, historic, and archeological sites; and the possible effects of environmental factors such as air, noise, and water pollution

4) Economic data including construction cost and the trend of industrial, agricultural, and commercial activities

2.3.2 Reconnaissance Survey

This phase aims at identifying several feasible routes, each within a group of a limited width of few hundred feet. When rural highways are proposed to construct, there is often limited information available on maps or photographs, and accordingly an aerial photography is used to obtain the required data. Feasible routes are identified by a stereoscopic examination of the aerial photographs, taking into consideration some factors such as:

1) Land and soil conditions
2) Serviceability of route to residential, industrial and commerci areas
3) Intersections of other transportation facilities, such as bridge railways, rivers, and other highways
4) Directness of route

2.3.3 Preliminary Location Survey

In this phase, preliminary alignments are used to evaluate the economic and environmental feasibility of the alternative routes. The purpose of economic evaluation is to assess each alternative route to determine the possible future impact on investing the resources required to financially support the construction the highway. Because the construction of a highway will possibly have impacts on its surrounding areas, the environmental evaluation is very important. The highway is therefore an integral part of the local environment and should be considered as such. This environment includes plant, and human communities and encompasses social, physical, natural, and man-made variables. These variables should interrelated in a manner that keeps equilibrium and sustains the lifestyle of the communities.

2.3.4 Final Location Survey

The final location survey is a detailed layout of the selected route. The horizontal and vertical alignments are determined and the positions of structures, and the drainage channels are located.

2.4 Factors Influencing Highway Design

Highways design is based on specified design standards and controls which depend on the following roadway system factors:

1) Functional classification
2) Design hourly traffic volume and vehicle mix
3) Design speed
4) Design vehicle
5) Cross section of the highway, such as lanes and shoulders.
6) Presence of heavy vehicles on steep grades
7) Topography of the area that the highway traverses
8) Level of service
9) Available funds
10) Safety
11) Social and environmental factors

These factors are often interrelated. For example, design speed depends on functional classification which is usually related to expected traffic volume. The design speed may also depend on the topography, particularly in cases where limited funds are available. In most instances, the principal factors used to determine the standards to which a particular highway will be designed are the level of service, expected traffic volume, design speed, and the design vehicle. These factors coupled with the basic characteristics of the driver, vehicle, and highway, are used to determine standards for the geometric characteristics of the highway, such as cross sections and horizontal and vertical alignments.

2.5 Highway Strata

Designers of highways should take into consideration many requirements including:

1) The design of highway should be enough to accommodate the expected traffic during the service life of the highway. This design may include number of lanes, horizontal and vertical curves and etc.

2) The structural design (pavement design) of highways should support the loads applied by traffic. This design may include numbers and type of strata (system of overlainanced layers) and provide adequate skid-resistant surface.

3) The design of highways should be cost-effective through considering the minimum maintenance.

Regarding the structural design of highways, selecting number and type of strata will relay not only on the expected traffic load but also on the in situ soil which the foundation of the design. However, the in-situ soil and the other overlying layers should ensure that the loads of traffic will distributed in such a way that none of the layers will be overstressed during the useful life of highways. In the structural design of the pavement, the followings should be identified:

The number of layers and the thickness of each layer.

1) The material to be used in each layer and the criteria of construction.
2) The expected applied loads of vehicles and other expected loads.
3) The volume of traffic during the life of the highway.
4) The soil properties of the in-situ soil and its bearing capacity.

The highway pavements normally consist of three layers and as follows:

1) Foundation: refer to the native sub-grade soil and the layer of graded stone (sub-base and possibly capping) overlaying the native soil. The sub-base and capping layer provide a platform on which the road-base material is placed. It also insulates the native soil layer from effects of inclement weather. These layers may act as a temporary road surface that is used during the construction phase of the highway.

2) Road-base: refer to the main structural layer that supposed to resist stresses and strains caused by vehicles' load. The road base should distribute stress in such a manner that the foundation layers do not become overstressed.

3) Surfacing: refers to top layers that in touch with vehicles tires . To ensure safe driving, the surfacing should have adequate skidding resistance. The surfacing should significantly reduce the probability of water infiltrating through surface cracks. Texture and durability are vital requirements of good pavement surface as are surface regularity and flexibility. For flexible pavements, the surfacing is normally applied in two layers: base-course and wearing course. The base course is an extension of the road-base layer providing a regulating course on which the final layer is applied. In the case of rigid pavements, the structural function of both the road-base and surfacing layers are integrated within the concrete slab.

2.6 Highway Drainage

Providing a sufficient drainage system is very important factor in the geometric design of highways. Drainage facilities in highways should be adequately designed for the flow of water away from the surface of the pavement to receivers. Inadequate design of the drainage system results in serious damages to the highway structure. Moreover, the traffic may be slowed by the accumulated water on the pavement, and accidents may occur as a result of hydroplaning and loss of visibility owing to splash and spray.

The importance of the drainage system is recognized in the amount of highway construction dollars allocated to drainage facilities. About 25 percent of highway construction dollars are spent for construction of drainage structures, such as culverts, bridges, channels, and ditches. The highway engineer is concerned primarily with two sources of water. First, surface water as rain or snow. Although some surface water is absorbed into the soil, the remaining surface water on the ground should be removed from the highway pavement.

Drainage of this source of water is referred to as a surface drainage. Second. ground water which flows in underground streams. This may become important in highway cuts or at locations where a high water table may exist near the pavement structure.

2.6.1 Highway Drainage Structures

Drainage structures are designed to carry traffic over natural waterways that flow below the right of way of the highway. These structures are also provided for the flow of water below the highway, along the natural channel, without significant alteration or disturbance to its normal course. The main concern of the highway designer is to provide an adequate size structure, such that the waterway opening is enough to discharge the expected flow of water. Inadequately sized drainage structures can result in water impounding, which leads to the failure of the adjacent sections of the highway.

The two general categories of drainage structures are major and minor drainage structures. Major structures are those with clear spans greater than 20 feet., while minor structures are those with clear spans of 20 feet or less. Major structures are large bridges, and multiple-span culverts, while minor structures include small bridges and culverts .

2.6.1.1 Major Structures

The emphasis is placed on selecting the span and vertical clearance requirements for such structures. The bridge deck should be located above the high water mark. The clearance above the high water mark depends on whether the waterway is navigable. If the waterway is navigable, the clearance above the high water mark should allow the largest ship using the channel to pass underneath the bridge without colliding with the bridge deck. The clearance height, type, and spacing of piers also depend on the probability of ice jams and the extent to which floating logs and debris appear on the waterway during high water.

6.1.2 Minor Structure

Minor structures, consisting of short-span bridges and culverts, are the predominant type of drainage structures on highways. Although openings for these structures are not designed to be adequate for the worst flood conditions, they should be large enough to accommodate the flow and conditions that might occur during the normal life expectancy of the structure. Provision also should be made for preventing clogging of the structure due to floating debris and large boulders rolling from the banks of steep channels. Culverts are made of different materials and in different shapes. Materials used to construct culverts include concrete (reinforced and unreinforced), corrugated steel, and corrugated aluminum.

2.7 Designations Associated with Highway Projects

In the terminology of the transportation engineering, readers can find many names to the pavement projects. This may include highway, freeway, street, road, and others, so the difference between one name and other should be explained. These differences are described below:

1) A beltway: (also called a loop) circles around a town or a city.

2) A causeway: is a roadway built on top of raised land made of compacted soil usually to cross a body of water.

3) A driveway: privately maintained access road leads to a single house or a group of houses.

4) A freeway: there are no traffic lights or intersections to slow traffic down. Traffic is usually divided using a median which are concrete barriers or sometimes grass.

5) A highway: refers to any public road that is well-maintained and can handle a heavy load of traffic.

6) The Interstate: The Interstate Highway System is a complex network of freeways that connect states together.

7) A parkway: is a major throughway originally designed for scenic and recreational driving.

8) A roundabout: is a circular intersection where traffic is designed to flow without stopping, in one direction.

9) A street: connects buildings together, while providing vehicles and pedestrians access. The distinction between *road* and *street* is that a road serves primarily vehicles and sometimes lacks a sidewalk, whereas streets are designed for pedestrian travel *and* vehicles.

2.8 Classification of Highway Projects

The network of roads and streets that connects and serves cities, towns, and villages is one of the most widely used means of transportation. In Iraq, as in many other countries, very high percent of all passenger travel is on roads and streets, and almost all of country's food and other goods are transported all or part of the way from the farm or factory to the store by way of roads and streets.

Highway can be classified into many groups in accordance with the following bases:

1) The location

2) The level of service

3) The upper portion of the pavement

2.8.1 Classification of Highways According to Location

In Iraq the paved roads carry the heaviest volume of traffic. They fall into six major categories: expressways, principal distributors, secondary distributors, collectors, local roads, and (military) road.

1) Expressways: They are main ways dedicated for vehicles with high speed design up to 120 km/hr characterized with high traffic volume up to 100000 vehicles per day on both sides. Moreover, they are long and without cross roads. These roads are characterized with the width of a single path, which is 3.75 m, and the width of the median, which is 3.0 to 8.0 m. In addition, they are without roadside and provided with safety fences and special path to stop vehicles in emergency cases.

2) Principal Distributors: These are the main roads that divide any city into many sectors. The essential function of these roads is to connect these sectors with each other. These roads are characterized by a high traffic volume up to 60000 vehicles daily on both sides. They permit the vehicles to move with speed designed up to 80 km/hr. Principal distributors

are long roads in which the width of a single path is 3.5 m, the median is 3.0 to 5.5 m. An additional path to vehicles stopping may be added with a width of 2.5 m, as well as, a pedestrian pavement of 2.0 to 10.0 m width.

3) Secondary Distributors: They are main roads dividing each sector of the city into main parts. Secondary distributors, from which local roads branch, are characterized with medium traffic density up to 40000 vehicles daily on both sides. Moreover, they permit the vehicles to move with a speed designed up to 60 km/hr. They differ from principal distributors in the sense they are shorter and similar in the width of the path which is 3.5 m and the median which is 2.5 m. A path vehicles stopping may be added with a width of 2.5 m, as well as, a pedestrian pavement which is 5-10 m.

4) Collector roads are urban and rural roads that connect cities and towns to the arterials (Principal Distributors) and interstates.

5) Local Roads: Local roads spread among buildings and houses, and they flow into secondary distributors. They are

characterized by a limited traffic volume, which does not exceed 10000 vehicles daily on both sides. These roads allow the movements of vehicles with a speed designed up to 40 km/hr. Local roads are often short and the width of a single path is 3 m without a median. A path vehicles stopping may be added of 1.5 m width, in addition, to the roadside which width does not exceed 5 m.

6) Military road mainly serves defense purposes.

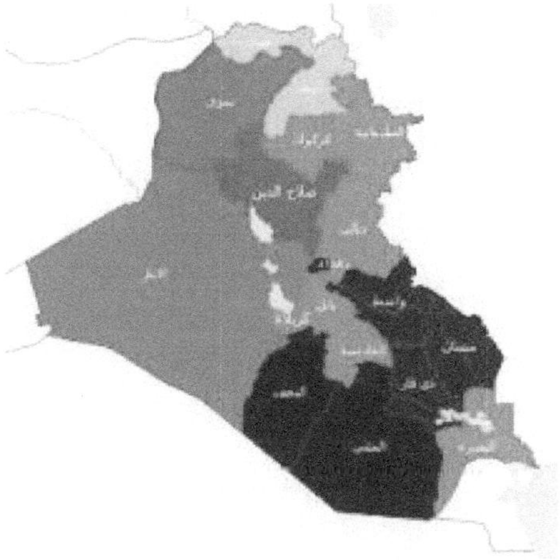

2.8.2 Classification of Highways According to Level of Service

Level of service (LOS) is a qualitative measure describing operational conditions within a traffic stream and their perception by motorists and /or passengers. Factors such as speed and travel time, freedom to maneuver, traffic interruptions, and comfort and conveniences are generally included as conditions affecting LOS. Each facility can be measured on the basis of six levels of service. A,B,C,D,E, and F. Level A representing the best operating conditions and level F the worst.

2.8.3 Classification of Highways According to Upper Portion of the Pavement.

There are many kinds of roads, such as elevated roads (bridges) and underpasses (tunnels). Most of these roads are stretched on the earth surface and they are classified into three kinds, flexible pavement, rigid pavement and composite pavement road.

1) Flexible Pavement: Flexible pavements have recently been developed into a wider variety of types than formally used. A conventional asphalt concrete pavement section, involves the use of an asphalt concrete surface layer, an asphalt concrete binder or leveling course layer (often of the same material as the surface layer but sometimes with a large maximum particle size), and one or more base and sub-base layer.

2) Rigid Pavement: There are four basic types of rigid pavements. these are derived from the combination of reinforcement and load-transfer devices within the concrete. Rigid pavements may be: Unreinforced, Lightly reinforced, Continuously, and Pre-stressed.

3) Composite Pavement: They are rigid paving roads to which a layer or more of asphalt concrete is added. The term composite pavement has been introduced from time to time but normally is not very effective definition. It is intended to describe a pavement that combines rigid and flexible element, such as a Portland cement concrete surface on an asphalt concrete base, or an asphalt concrete surface (usually an overlay) over an old Portland cement concrete (rigid) base or over a Portland cement treated (rigid) base. A more useful definition is to assign this type of pavement to one of the other two types, depending on the basic load-carrying element and not the visible surface type.

2.9 Highway Project Life Cycle

The highway project construction process can be divided into five phases. these are:

1) Decision Phase,
2) Design Phase,
3) Construction Phase,
4) Commissioning Phase and
5) Maintenance Phase.

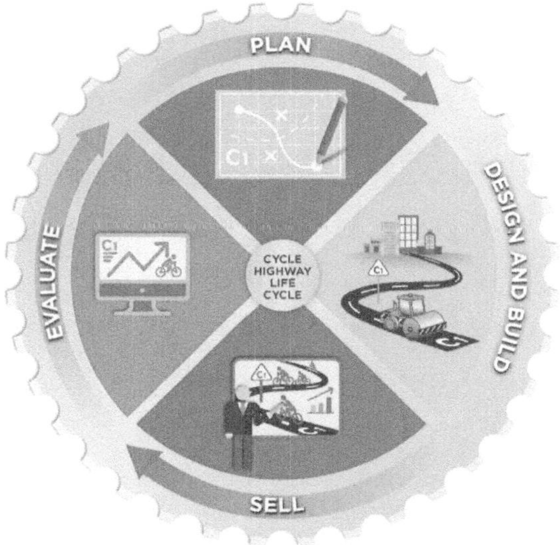

2.9.1 Decision Phase

The first phase of a project is the decision phase. During this phase, a highway-engineering project begins with a thorough investigation of its technical and economic feasibility. The characteristics of highway engineering projects require extensive investigations, such as geological survey, so that alternative solutions may be evaluated sensibly. Even so, the final outcome of these investigations is often that several, perhaps widely different, solutions are equally attractive, and it may be difficult to weigh the advantages of each scheme, particularly in the long-term look.

Thus the outcome of this stage of highway engineering project is the "report", which describes the investigations and analyses that have been performed and also describes the possible solutions and evaluates them. The report should also include recommendations to help the client in making his choice to start with the project. The main purpose of this stage is:

1) To enable an objective decision to be made on the feasibility of the project, and to examine the possible alternatives for the client to make his decision of proceeding with the project.

2) To specify the client's requirements in detail so that the designers may correctly interpret his wishes and estimate the likely costs.

2.9.2 Design Phase

Once the scope of the project has been defined in the terms of reference, the project enters the design phase. The main purpose of this stage is to:

1) Design the highway project in detail.
2) Obtain the necessary approvals.
3) Estimate the pre-construction costs.
4) Prepare the necessary production information.

The activities at this stage will include:

1) Deciding which solution to adopt regarding alignment, grades, sight distance, lane and roadway width, surfacing, right-of-way width, median width, rail crossing, roadway structures, etc

2) Undertaking further investigations, if necessary.

3) Some items must also be prepared such as:

    a) A scheme design, including cost estimate;
    b) A detailed design;
    c) Working drawings, specifications and schedules;
    d) Bill of quantities;
    e) Cost estimates; and
    f) A preliminary construction programmed, including time-schedule.

The design stage includes three managerial stages:

1) Design study (Value Engineering)
2) Pre-planning (project common time schedule)
3) Estimating and tendering.

The purpose of estimating and tendering is to appoint a highway-engineering contractor, or a number of highway engineering and other contractors, who will undertake the site construction work. It is aimed to obtain tenders from contractors for the construction of the works and to award the contract.

2.9.3 Construction Phase

The main purpose of the construction stage is to construct the highway engineering projects within the limits of cost and time, and specified quality. This phase can be divided into two stages:

1) Site Preparation.
2) Construction Works.

2.9.3.1 Site Preparation.

This denotes clearing and grubbing operations which include removal from the construction area of; trees, underbrush, stumps, rubbish, abandoned drainage structures and sometimes other obstructions such as buildings, fences, utility lines, etc.

It may be desirable to preserve tree, shrubs, etc. growing within road limits. Such trees should be carefully marked and, if necessary, protected against accidental damage during construction by putting tree guards around them. All waste material, if combustible, should be removed from road to waste areas.

After the site is prepared, it should be inspected for satisfactory condition before earthwork is started otherwise organic material or debris intended for removal may get buried.

2.9.3.2 Construction Works

There are two main types of highway pavement: flexible and rigid pavements. The fundamental steps in constructing flexible pavement may be listed as follows:

1) Preparation of mixture: the preparation of the pavement mixture may be either by central mixing plants or batch type.

2) Transportation of mixture: the mixture is discharged from the plant into dump trucks or trailers for transportation to the job site.

3) Preparation of the base: surface courses of asphaltic concrete are frequently placed on new or existing bases that require very little preparation prior to the placing of the new surface other than through sweeping and cleaning to remove loose dirt and other foreign material.

4) Placing of mixture: placing of the asphaltic concrete mixture is permitted only when the base is dry and under favorable weather conditions.

5) Compaction of the mixture: when the spreading and finishing operations have been completed, and while the mixture is still hot, rolling begins. Rolling procedures vary with the properties of the mixture, thickness of layer, and other factors.

The fundamental steps in construction rigid pavement may be listed as follows:

1) Preparation and preliminary finishing of the sub-grade: the preparation of the sub-grade on which a concrete pavement is, of course, very important step in the overall construction process. It is essential that uniform sub-grade support is provided for the completed pavement during its service life and that the pavement be free from other detrimental effects associated with unsatisfactory sub-grade soils. In some cases the existing sub-grade soil may be entirely suitable as a foundation and thus requires only compaction and adequate drainage prior to the placing of a concrete slab. In other instances a layer of imported soil may be used between the existing soil and the slab, or the existing soil may be stabilized, as with Portland cement, bituminous materials, or granular materials. It is a common practice to provide a blanket or "installation" course of suitable granular material between the existing soil and the pavement itself.

Various materials are used to form this layer or "sub-grade" including gravel, sand and gravel, crushed rock, and so on. The depth of treatment varies from a minimum of 8 to 16 cm. or so if the primary purpose is to prevent pumping, vary much thicker layers of granular materials may be used to prevent frost action in northern parts of the country.

2) Placing of forms: most concrete highway pavement is constructed with the slip form pavers.

3) The final shaping of the sub-grade to the exact dimensions required by the plans and specifications. This operation is generally accomplished by a machine called a "sub-grader".

4) Installation of joints: the installation of the various types of joints that may be used in a concrete pavement is also a very important step in the construction process. A portion of the process of constructing the required joints, which we may call the "the installation of joint assemblies", normally takes place between the final finishing of the sub-grade and be beginning of actual concrete operations.

5) Mixing of concrete: at this point of the construction of concrete pavements, all preparation has been made at the site for the placing of concrete between the forms. Formally, aggregates and cement were batched "dry" at a convenient location some distance away from the job site, brought to the site in trucks, and mixed with water on the job. Central mixing plants vary widely. In urban areas, the plant often is a highly sophisticated, stationary "ready-mix" plant capable of producing automatically many mixtures of varying composition for many purposes. Other plants are portable or semi-portable; they are automatically controlled and can be moved from one location to another quite easily.

6) Placing and finishing: placing of the fresh concrete at the job site and finishing the job to required smoothness are important operations that are done in a variety of ways.

7) Curing: curing of a concrete paving slab is necessary in order that the concrete hardens proper.

The construction stage also includes the following managerial activities:

1) Construction planning, which includes the formulation of:
   a) Time schedules;
   b) Site organization;
   c) A manpower plan;
   d) A material procurement plan.

2) Co-ordinate and control site operations, which includes:
   a) All temporary and permanent construction works, and the supply of all built-in furniture and equipment;
   b) The co-ordination of subcontractors;
   c) General supervision;
   d) Site costing and cost monitoring.

The construction stage of a highway consists of a number of inter-related activities. The failure of one activity can disrupt the entire production schedule. Therefore, careful production planning is important.

The construction stage is usually the most expensive, intensive and difficult of the highway project stages.

2.9.4 Commissioning Phase

The commissioning stage in highway–engineering project is essentially concerned with testing the engineering performance of the works. Works that are buried such as the base and sub-base course for example are usually tested before they are covered. Other works can be tested only when completed. Traffic signals are examples of these.

The commissioning activities for most highway engineering projects will include the following:

1) Testing the engineering performance and safety of the entire work package;

2) Searching for non-operational defects, (e.g. those that affect only the appearance);

3) Remedying all deficiencies;

4) Preparing "as-built" drawings and other records;

5) Preparing operational instructions and maintenance manual;

6) Training staff;

7) Monitoring performance of works against original requirement.

The main purpose of this stage is :

1) To ensure that the highway engineering works have been completed as specified in the contract documents, and that all facilities work is properly made.

2) To provide a record of actual construction, together with operating instruction.

3) To train staff in the use of works.

2.9.5 Usage (Maintenance) Phase

Highway maintenance is defined as "the function to preserve, repair and strengthen the highway, and keeping it in condition for safe, convenient and economical use". Normal maintenance operations consist of the following general items (activities):

1) Surface repairs and upkeep;
2) Shoulder and sub-grade maintenance;
3) Repair, upkeep and cleaning of drainage facilities, including ditches and culverts;
4) Roadside maintenance, including repair and control of erosion mowing of slopes, control of obnoxious weeds, and maintenance of footpaths, recreation areas;
5) Snow and ice removal and control, and street cleaning;
6) Bridge repair , painting and maintenance of stream beds against hazardous erosion or filling up;
7) Traffic service, including the maintenance and installation of signs, painting of traffic stripes, signal-and traffic-light maintenance and operation, and guard-rail painting and repair;
8) Special service, such as the issuance of permits for heavily loaded or over-sized vehicles, permit for street and highway openings, and permit for work within or abutting the roadway by property owners when such work affects the highway facility.

CHAPTER THREE

3 PAVEMENT DETERIORATION

3.1 Introduction

The deterioration of a pavement can be distinguished by various external markers or indicators, and often these indicators reveal the probable causes of the failure. The discussions of problems regarded to pavement distress generally depend on whether the pavement has a flexible or rigid surface type.

3.2 Modes of Pavement Distresses

As previously mentioned that external markers and indicators may reveal the probable causes of failures. Generally, the various types of distresses fall into one of the following categories:

1) Cracking
2) Joint Seal Damage
3) Disintegration
4) Distortion
5) Loss of skid resistance

3.2.1 Modes of Rigid Pavement Distresses

Cracks occurred in rigid pavement can be classified into hairy, medium and wide cracks based on their width. These cracks can be longitudinal or transverse. Some of these cracks can be observed immediately after drying out of concrete, others can be generated during the service life of the pavement. Hairy cracks refer to those less than 0.5 mm wide while wide cracks refer to those more than 15 mm. Regardless the wideness of the cracks, they are paths to infiltrate the surface water to the underlying layers leading to fail of the layers. Wide cracks may result in loosing of the aggregate interlocks and deterioration of pavement structure. Use of reinforcement in concrete pavement can limit the generation of the different cracks. For plain concrete, the failure can be generated as a result of one of the following:

1) Presence of medium and/or wide cracks (longitudinal c transverse) crossing the bay of the concrete slab.
2) Presence of medium intersecting longitudinal transverse crack: The length of these cracks are more than 200 mm in an starting from the corner of the pavement
3) Presence of wide corner cracks. These are centered on th concrete corner and are more than 200 mm in radius.

a. Cracking

In rigid pavements, cracks often result from stresses resulted by expansion and contraction of the pavement. Insufficient and/or improperly cut joints acting separately or in combination may lead to cracking. There are different types of cracking that can be occurred including:

1) Longitudinal, diagonal, and transverse Cracks
2) Breaks of corner
3) Durability "D" Cracking
4) Cracks owing to shrinkage
5) Damage owing to joint seal.

b. Disintegration

Disintegration is the breakage of a pavement into small parts combined by loose particles. Dislodging of aggregate particles is also considered as disintegration. Unsuitable aggregates, poor mixing of concrete pavement, and improper curing and finishing of the concrete can cause the disintegration. This defect falls into several types:

1) Scaling
2) sapling
3) Blow up and Pop-outs
4) Patching.

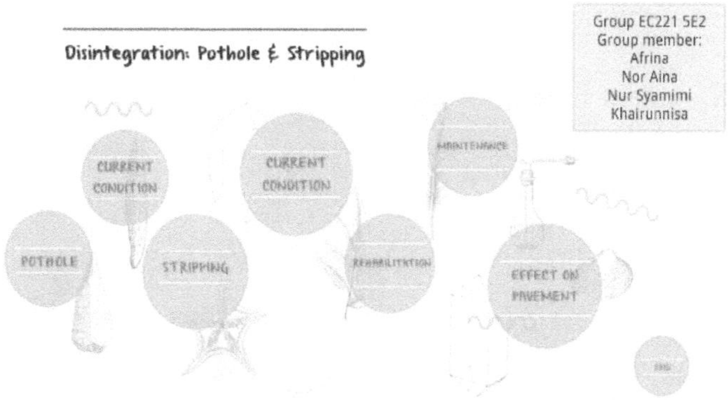

c. Distortion

Distortion is a change in the original position of pavement surface. Distortion can be resulted from expansive soils, loss of fines, or foundation settlement owing to improperly designed drainage systems. Generally, there are two types of distortion:

1) Pumping
2) Settlement or Faulting.

d. Loss of Skid Resistance

Skid resistance is the ability of a pavement to provide the required friction characteristics for the road surface under all weather conditions. Also, it is a function of the surface texture. The loss of skid resistance results from the textured surface wearing down through regular wear and tear or owing to the accumulation of contaminants.

e. Polished Aggregates

Some particles of aggregates become polished quickly under traffi load. Natural polished aggregates may create skid hazards if use alone in the pavement without crushing aggregates. Crushin aggregates provide rough angular faces that enhance skid resistance.

3.2.2 Modes of Flexible Pavement Distresses

For heavily trafficked roads, experience has indicated that deterioration in the form of cracking/deformation is most likely to be occurred in the surface of the pavement rather than deeper down within its structure. A well-constructed pavement will have an extended life span on condition that distress, seen in the form of surface cracks and ruts, is taken care of before affecting the structural integrity of the highway. The structural deterioration of flexible pavements passes through four main phases as listed below:

1) **Phase 1**: During gaining of strength and reaching stability. In other words, the load spreading ability of the pavement is still improving.
2) **Phase 2**: The ability of load transferring of the pavement is uniform. At this phase, the rate of structural deterioration can be predicted with some confidence.
3) **Phase 3**: This is an investigatory phase where defects and deterioration are difficult to predict because of the gradual or even rapidly reduction in strength of the pavement. The need of the possible remedial action should be determined in this phase through monitoring of the pavement.

4) **Phase 4**: In this phase, the deterioration of the pavement reaches to failure. Reconstruction is the only way to get back the strength of the pavement. This phase may take many years as the repeated maintenance may delay the reconstruction. The increasing frequency of maintenance with associated heavy costs make reconstruction required and cost-effective.

It is worthy stating that the remedial actions that carried out during third phase "investigatory" is absolutely more economic than complete reconstruction of the pavement.

a. Cracking

Cracks in flexible pavements result from the deflection of surface over an unstable foundation, thermal expansion and contraction of the surface, shrinkage of the surface, reflection cracking, or poorly constructed lane joints. In flexible pavement, five types of cracks commonly occur as follows:

1) Longitudinal and Transverse Cracks.
2) Block Cracking
3) Reflection Cracking
4) Alligator or Fatigue Cracking
5) Slippage Cracks.

b. Disintegration

Many causes lead to disintegration in a flexible pavement. These causes include insufficient compaction of the surface, severe overheating of the mix, adhesion loss between the asphalt coating and aggregate, or insufficient asphalt binder in the mix. The following are common types of disintegration:

1) Raveling and Weathering
2) Potholes
3) Asphalt Stripping
4) Jet Blast Erosion
5) Patching and Utility Cut Patch.

c. Distortion

Distortion is caused by foundation settlement, lack of stability in the bituminous mix, insufficient compaction of the pavement courses, poor bond between the underlying layer of pavement structure and the surface, and swelling subgrade soils. The following are common types of distortion:

1) Rutting
2) Corrugation
3) Shoving
4) Depression
5) Swelling.

d. Loss of Skid Resistance

The decreasing of skid resistance for a pavement surface is caused by very heavy tack coat, poor aggregate subject to wear, high content asphalt in the mix of bituminous, and contaminants buildup. A loss of skid resistance in a pavement surface may result from the following:

1) Polished Aggregate

2) Contaminants

3) Bleeding

4) Fuel/Oil Spillage.

As previously mentioned, there are two types of pavements so that the deterioration in one type will be different from another. Each type of deterioration will be separately described in the next sections.

3.3 Causes of Pavement Distresses

Pavement are designed according to different specifications owing to the differences in the environment, resources and purposes. However, pavement are more subject to damage due to change in friction and hitting powers between the vehicle's wheels and the surface of the road. Moreover, roads are exposed to nature factors, which play a damaging role, resulting from variation of temperature and humidity between day and night or from one season to another.

Damage in the pavement happens as results of one or more of the following reasons:

1) Natural and weather factors.

2) Mistakes in the design.

3) Mistakes in the construction methods.

4) Failure in the quality control of materials.

5) Negligence of periodic maintenance.

6) Misuse of roads such as exceeding the allowed limits.

There are three primary causes that lead to the occurrence of the distresses in pavement. These causes are traffic loading, environmental damage, and inadequate construction methods, design, or material.

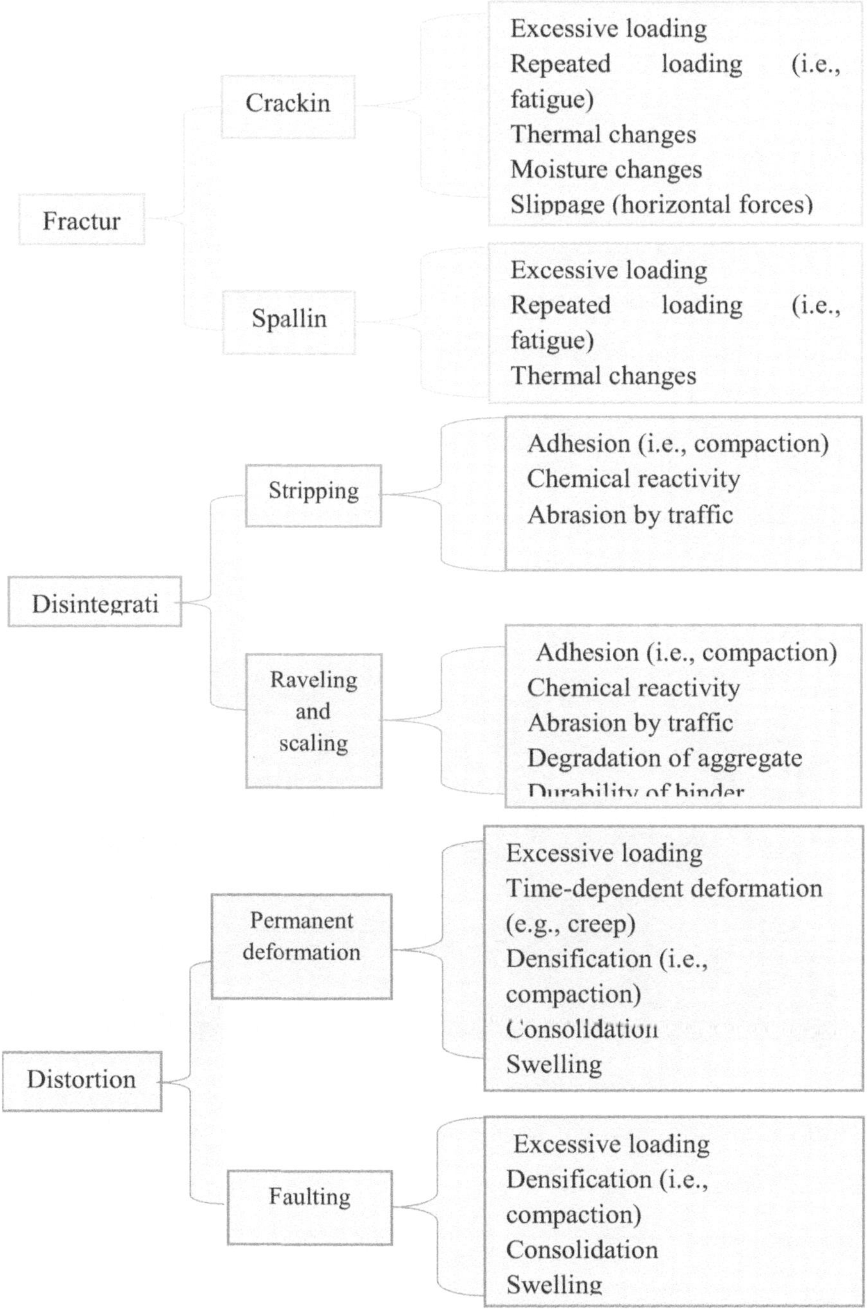

Figure (3.1): Causes of Pavement Distresses.

Generally, the various types of distresses fall into one of the following categories:

3.4 Rutting

Description: A rut is a surface depression in the wheel tracks. This can be combined by pavement uplift along the sides of the rut. In many instances ruts can be noticed only after a rainfall when the paths are filled with water. Rutting is owing to a permanent deformation in any of the pavement layers or subgrades. It is usually caused by consolidated or lateral movement of the materials owing to traffic loads. Significant rutting may lead to major structural defects of the pavement.

Figure (3.2): Rutting Distress.

Rutting is a longitudinal surface depression in the wheel tracks, and may represents serious hazard for vehicle movement. For low to medium severity rutting, a partial depth patch can be used to recover the pavement surface. Higher severity rutting may require a mill and overlay or even reconstruction.

3.5 Depressions

Description: localized areas in the pavement surface with slightl lower elevations than those of surrounding pavement. In many case: light depressions are not visible until after rain when ponding wate creates a "bird bath" areas on pavement. Stains caused by pondin water are the main indication of depression. Settlement of th foundation soil or an improper construction can be the main reason of depressions.

Figure (3.3): Depression Distress.

Depressions are normally combined by some roughnes: Depressions cause hydroplaning are deep or filled with water. Sag: unlike depressions, are abrupt drops in elevation. A depression in th pavement is sometimes referred to as a birdbath, because it ma collect water. Low severity depressions may not need repairs, bu those that cause more ponding must be treated (remove and replace)

3.6 Corrugation

Description: Corrugation ,also known as wash boarding, is a chain of closely spaced ridges and valleys (ripples). It occurs at fairly regular intervals, usually less than 3m (10 ft) along the flexible pavement. The ridges and valleys are perpendicular to the traffic direction. This type of distress is usually owing to traffic action combined with an improper pavement surface or base. If bumps be occurred in series of less than 3m regardless any cause, the distress is considered as a corrugation.

Figure (3.4): Corrugation Distress.

This distress is also called "wash boarding" because the pavement surface looks similar to a washboard. In other words, corrugation is closely spaced transverse bumps perpendicular to traffic direction. It usually occurs at road intersections where traffic is starting and stopping. No repair is required for low severity corrugations, but patching (partial or full depth) of the surface layer, or a mill and overlay, may be needed to repair medium to high severity distresses.

3.7 Potholes

Description: Potholes are small shaped depressions in th pavement surface. They are usually less than 0.9m in diamete Potholes have vertical sides and sharp edges sides near the top of th hole. The growth of Potholes is accelerated by collecting of fre moisture inside the hole. Potholes are created when traffic scrape small particles of the pavement surface. Then, the pavemen continues to degraded owing to the weak spots in the surface an poor surface mixtures. Potholes can be occurred when pavement ha been reached a condition of high-severity alligator cracking. Althoug raveling and weathering can accelerate the severity, Potholes ar structurally related distresses . When holes are created owing to higl severity alligator cracking, they should be classified as potholes not a weathering.

Figure (3.5): Potholes Distress.

Insufficient pavement strength can result in Patholes. Dependin on the severity of the Potholes, the pavement may be partially or full removed.

3.8 Shoving and Pushing

Description: Shoving is defined as a permanent, longitudinal displacement of localized areas in the pavement surface. It mainly caused by the traffic loading. The traffic pushes against the pavement producing short and rapid waves in the pavement surface. This distress normally occurs only in improper liquid asphalt mix (cut back or emulsion) pavement.

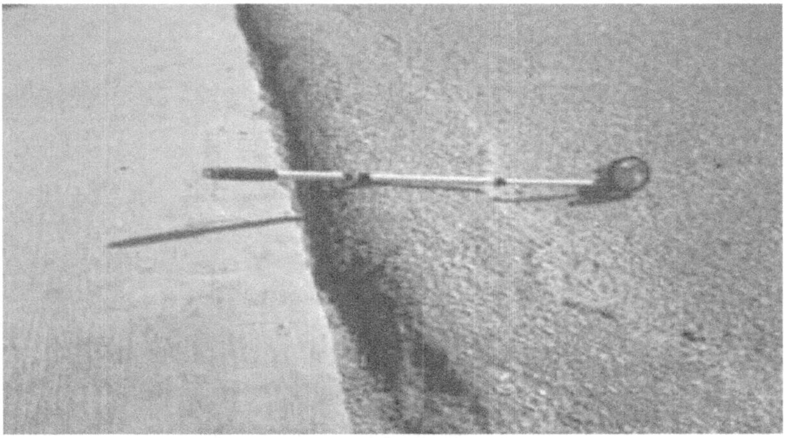

Figure (3.6): Shoving Distress.

Shoving as shown in Figure 3.6 is a movement of a localized pavement areas owing to traffic loading. It is a movement parallel to the direction of traffic. The treatment can be ranged from no repair for low severity shoving to full depth patching for more severe shoving.

3.9 Weathering and Raveling

Description: Weathering and raveling are the wearing away of th pavement surface due to a loss of asphalt or tar binder and dislodge aggregate particles. These distresses indicate that either the aspha binder has hardened appreciably or a poor-quality mixture is presen In addition, raveling may be caused by certain types of traffic, e.g tracked vehicles. Softening of the surface and dislodging of th aggregates due to oil spillage are also included under raveling.

Figure (3.7): Raveling Distress.

Raveling is the disintegration of the pavement from the surfac down owing to loss of asphalt binder and dislodged aggregate piece: No repair is needed for low severity raveling while an overlay may b required to correct higher severity distress.

3.10 Polished Aggregate

Description: Repeated traffic application may cause this kind of distress. When the aggregates in the surface become smooth touch, friction with vehicle tires is significantly decreased. In other words, when the portion of aggregate above the surface (in touch with tires) becomes very smooth, the texture of the pavement does not really contribute in reducing of the vehicle speed. If close examination reveals that the aggregate above the asphalt is negligible, and the surface of aggregate is smooth, the distress of polished aggregates should be taken into account. Test of skid resistance should be performed and compared with previous tests to have an indication of this kind of distress.

Figure (3.8): Polished Aggregate Distress.

Polished aggregate occurs when the pavement surface becomes smooth to the touch. This leads to a low skid resistance. To restore the surface friction an overlay, pavement grooving, or grinding may be required.

3.11 Bleeding or Excess Asphalt

Description: Bleeding can be easily distinguished as a film c bituminous material on the pavement surface. It creates a shin glasslike surface that usually becomes quite sticky. Bleeding is mainl attributed to the use of excessive asphalt cement or tars in the mi The excessive applications of bituminous sealant, and/or low air voi content are also causes of this kind of distress. It occurs when aspha fills up the voids in the mix during hot weather and then bleeds t the pavement surface. Asphalt or tar will accumulate on the surfac as the bleeding process is not reversible during cold weather,.

Figure (3.9): Bleeding Distress.

Bleeding, which sometimes is called flushing, is owing to th unstable asphalt mixture. Consequently, traffic loadings cause th asphalt in the pavement to bleed up over the surface. No repair i required in low severity bleeding, but pavements with high severit bleeding may lose their skid resistance and should be milled an overlaid.

3.12 Bumps and Sage

Description: Bumps are upward, small, localized, movement of the pavement surface. They are different from shores in the sense that shores are owing to unstable pavements while bumpscan be caused by several factors, including:

a) Buckling or bulging of underlying PCC slab.

b) Frost heave (ice lens growth)

c) Infiltration and buildup of material in a cracks of pavement surface in combination with traffic loading "sometimes called a tenting".

Sage are small, abrupt, downward movements of the pavement surface. Swelling on the other hand are displacement that occur over large areas of the pavement surface.

Figure (3.10): Pumping Distress.

3.13 Swell

Description: swell is defined as an upward bulge in th pavement's surface-along,. Swelling can be accompanied by surfac cracking and this distress is usually owing to frost action in the su grade or because the present of swelling soil.

Figure (3.11): Swelling Distress.

Swelling is an upward bulge in the pavement surface ma result in surface cracking. Low severity swelling may settle when th weather warms and therefore no repair is required. Full dept patching with sub-grade correction may be needed for medium t high severity of this distress.

3.14 Cracks

There are two main types of cracks: transverse cracks and longitudinal cracks.

3.14.1 Transverses cracks

Transverse crack extends across the pavement at approximately right angles to the pavement centerline or direction of lay down. These types of cracks are not usually loads associated.

Joint sealant Materials can be used to repair these cracks. Full depth treatment may be required for more sever transverse cracks which sometimes combined by secondary cracking, indicating loss of support in the underlying base layer.

Figure (3.12): Transverse Cracking Distress

3.14.2 Longitudinal Cracks

Longitudinal cracks are parallel to the pavement's centerline c the lay down direction. They may be caused by:

a) A poorly constructed paving lane joint.

b) A shrinkage of the AC surface due to low temperatures or hardening of the asphalt and/or daily-temperature cycling.

c) Reflective cracks caused by cracking beneath the surface, including cracks in PCC slabs. These cracks may or may not be related to traffic loads. Cracks should be patched or sealed depending on the severity.

There are several types of longitudinal cracks.

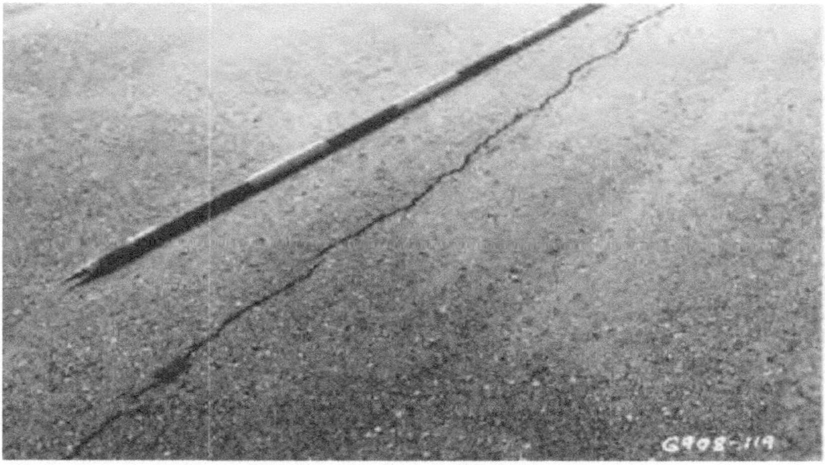

Figure (3.13): Longitudinal Cracking Distress.

3.14.2.1 Edge joint cracks

Description: Edge cracks are usaually parallel to and usually within (2ft or 0.6m) of the outer edge of the pavement. This distress is accelerated by traffic loading and can be caused by frost-weakened base or sub grade. The area between the cracks and pavement edge, is classified as raveled if it breaks up (sometimes to the extent that pieces are removed).

Figure (3.14): Edge Cracking Distress.

Edge cracks are within 1-2 feet of the outer edge of the pavement surface. These type of cracks may create alligator cracking along pavement edge. Repair may range from using sealant to patching the entire pavement depending on the severity of cracking.

3.14.2.2 Lane joint cracks

These cracks are located between road lanes and caused by th weakness pound between the distribution's lanes or the traffi loading. Traffic loading on the one lane is different from the othe lane.

Figure (3.15): Lane/Shoulder Drop off Distress

This distress can be noticed as a difference in elevation betwee shoulder and pavement edge and. This distress is attributed t shoulder erosion, shoulder settlement. It should be noted tha building up the roadway without adjusting the shoulder level ca cause this kind of distress. No repair is required for low severity, bu for medium to high severity, re-grading and filling shoulders to matc pavement height is required.

3.14.2.3 Alligator cracks

Alligator, sometimes called as fatigue cracking, is a series interconnecting cracks. This type of cracks is caused by repeated traffic loading on the asphalt concrete surface (fatigue failure). Cracking begins at the bottom of the asphalt surface or stabilized base where tensile stress and strain are the highest under a wheel load. These cracks looks like an alligator's skin. Alligator cracking is considered as a major structural distress and is usually accompanied by rutting as shown in Figure 3.16.

Figure (3.16): Alligator Distress.

Removing of the pavement with patching or reconstruction of the pavement may be the only method to repair the alligator cracking.

3.14.2.4 Shrinkage cracks

Shrinkage or "block" cracks are interconnected and intersecte cracks that divide the pavement into rectangular or square piece: The blocks may range in size from approximately (0.3mx0.3m.) t (3mx3m). Shrinkage cracking is mainly attributed to the shrinkage c the asphalt concrete and day/night temperature cycling (whic creates daily stress/strain cycling). This kind of cracks are not loac associated. Block cracking usually gives an indication that the aspha has been significantly hardened. Shrinkage cracking normally spreac over a large pavement area. Sometimes, block cracking also occur i non-traffic area. This type of distress can be distinguished fror alligator cracking in that the alligator cracks have many-sided piece with sharp angles. Also, unlike block, alligator cracks are caused b (fatigue) repeated traffic loading, and are therefore present in highl loaded traffic area such as wheel tracks.

Figure (3.17): Block Cracking Distress.

3.14.2.5 Slippage cracking

Slippage cracks are crescent (half-moon) shaped cracks. These cracks are mainly seen in areas of braking or turning wheels causing the pavement surface to deform and slide. This distress usually occurs when the surface mix is low in strength or the bond between the surface and the next layer of the pavement is poor.

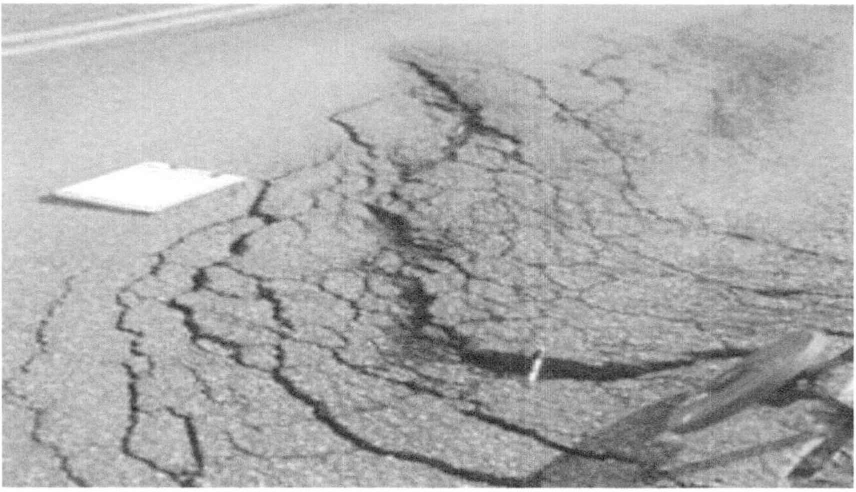

Figure (3.18): Slippage Cracking Distress.

This distress is a crescent or moon shaped, and owing to the low stability of the asphalt mixture or to a poor bond between layers of pavement layers. Reconstruction or patching may be required to repair high severity slippage cracks while no repair is required for low severity slippage cracking.

3.14.2.6 Reflection cracks

Cracks appear on a surface or overlay that are caused by joints c cracks in underlining. They may also be described as longitudinal c transverse in relation to the road centerline. Reflection cracks aı created because of overlaying above underlying joints or cracks. Th pavement may be spelled along the crack. While these cracks can b repaired by joint sealant materials in case of low severity cracks, th entire surface should be removed and the underlying cracks to b repaired prior to overlying in more severe cases.

3.14.2.7 Corner Break

Pavements with this distress have a corner of the slab broken in triangular piece. No repair is required for low severity corner break: but crack sealing or full depth patching may be required for slabs c the worse condition.

Figure (3.20): Corner Break Distress.

3.14.2.8 Block Cracking

These two way cracks divide the pavement into blocks approximately 1 to 12 feet square. Seal coats or thin overlay can be used to repair these kind of cracks.

Figure (3.21): Block Cracking Distress.

3.14.2.9 Durability Cracking

This distress has a pattern of running parallel cracks and it locates close to joints. It looks like a series of fine, hairline cracks and usually moves across the slab corners. This type of crack sometimes results in a disintegration of the entire slab. No repair is required with low severity distress while partial or full depth patching should be used on pavement in worse condition. Total slab replacement may be considered on pavements with extreme severity durability cracking.

Figure (3.22): Durability Cracking Distress.

3.15 Punchouts

A punch-out is a localized area of the slab that has been broken int pieces. No repair is required for low severity punchouts, while sealin or slab replacement can be considered for more severely damage pavements by punchouts.

Figure (3.23): Punchouts Distress.

3.16 Patching

A patch is considered a defect, regardless of how well it is performing.

Figure (3.24): Patching Distress

CHAPTER FOUR

4 HIGHY WAY MAINTENANCE OPERATIONS

4.1 Introduction

Maintenance, as defined earlier, involves keeping facilities as near the "constructed or reconstructed" condition as possible depending on circumstances. This kind of maintenance requires a periodic or almost immediate correction of unfavorable or unacceptable situations. Maintenance that will be discussed in this chapter will include the maintenance of the pavement surface, shoulders and approaches, road sides and drainage, the snow and ice control, bridges and traffic service

4.2 Pavement Surface Maintenance

Day to day assurance of roadway surface integrity is a primary responsibility of maintenance. Roadway surface includes the travel way and shoulders. Additional requirements for shoulders maintenance are presented in section 4.3. It is estimated that about one-half of the state highway maintenance money return goes for routine care of the roadway surface.

Figure (4.1): Operation of Surface Maintenance.

Travel ways are structured of two parts, described as:

Pavement structure is made up of layers described as sub-base, base coarse and surface coarse, and

Subgrade, which is the graded portion of a highway prepared as a foundation for the travel way pavement structure and shoulders.

Surface courses may be untreated or treated with admixtures or may be paved. Pavement structure is usually described as either flexible or rigid for bituminous type and Portland cement concrete type, respectively.

Maintenance requirements for roadway surface are intended to offset the effects of edge, wear, and various type of distress. The type of distress and the maintenance method employed are largely influenced by the type of the pavement structure and materials used.

Factors influenced the life of a roadway surface may include: temperature change, moisture content, structural material, design and the type and amount of traffic. The quality of maintenance is a considerable factor in preserving the lifetime of the pavement surface and its serviceability. The quality of maintenance is based upon a combination of using proper materials and the right time of use

Maintenance programs of asphaltic pavement surfaces are divided into two categories based on selecting the right materials and employing the best available methods. The first category includes operations such as patching, crack filling and use of surface treatments such as fog and sand seals. The second category includes resurfacing operations that involve precise determination of material characteristics and quantities to assure proper bonding.

Under the second category, the technology required to produce an effective surface by slurry seal or overlay should not be underestimated. Taking into consideration the time and materials that can be invested, it is better to use products with best expected results. Factors that affect the resurfacing are, type of roadway, shape and condition, amount of grade and quality of aggregates, ambient air and materials temperature; degree of surface moisture and humidity; condition and preparation of surfaces to be treated; traffic volume; equipment and method used; and grades of asphalt and aggregate and their relation to other materials, methods and carrying time. Materials tested in labs can prevent losses when sizeable amounts of paving materials are maintained. Many surface failures are owing to either improper construction or design. Determining the cause of failures is essential to proper maintenance and repair. Failures caused by design and those due to increasingly heavy traffic should refer to design engineers for correction construction design. Surface failures that are attributed to poor construction usually require increasing amounts of corrective action by maintenance over the life span of the pavement.

4.3 Maintenance of Shoulders and Approaches

Maintenance requirements of shoulders and approaches are influenced by their design, usage, condition and structural materials. Shoulders are parallel and contiguous to the travel way to which they provide lateral support. Shoulders should be suitable to provide emergency parking or serve as deceleration lanes.

Figure (4.2): Shoulders Maintenance Operation.

Shoulders should be designed to prevent collection of water in pockets and to assure good drainage The Shoulder width and cross slopes are determined by design standards on new highways. It should always maintain to proper grade.

Newly constructed shoulders should be checked frequently during the rainy season for damage, particularly for any slumping that may affect the shoulder's ability to provide lateral support to the travel way. There are four types of shoulders:

4.3.1 Earth or Sod

Earth or sodded shoulders should contain a sufficient amount of granular material for stability. Grass cover should be natural and, not requiring special irrigation. Earth or sod will tend to build up and should be bladed lightly to maintain proper cross slope drainage. Blading should be performed mostly when the earth is moist, often enough to keep ruts filled and lightly enough to prevent damage to grass roots. Ruts and soft spots appear after a rain or when ground frost has thawed and attention at these critical times should be taken. If the shoulder material has been removed by erosion, it should be replaced. In regions of heavy growth, use of a disc harrow or mechanical mixer in the early spring to break up roots and sprigs followed by blading is recommended. This will assist the normal growth of grass on the shoulders.

Figure (4.3): Maintenance Operation of Earth Shoulders.

Sod shoulders should not be mowed shorter than (3″-4″) or the effectiveness of the grass cover will be lost.

4.3.2 Gravel or crushed stone

Gravel shoulders offer better support for traffic than sod and it is more desirable as an all-weather shoulder. When a smooth surface or a proper slope is required, the work can best be accomplished by a motor grader. The good procedure is to blade the gravel to the edge of the pavement, and to move it back over the full width of the shoulder on the second pass.

Care should be taken not to leave ridges or false ditches to trap water on the shoulder. Dirt or debris from the slope or ditch area should not be mixed with shoulder gravel. Immediately following the blading of the gravel shoulder, compaction should be applied.

Figure (4.4): Maintenance Operation of Gravel_Shoulders.

The application of dust palliatives on some shoulders may be advisable during the dry season particularly if there is a high traffic volume. Herbicides to prevent the vegetation growth may be required. They should be applied prior to the growing season.

4.3.3 Bituminous Pavement

Paved shoulders with bituminous offer better and safer facilities for acceleration and deceleration of vehicles. Paved shoulders should be patched and sealed as described in item "roadway surfaces" . Sealing cracks and joints in shoulder pavement should be treated as prescribed for similar types of roadway surfaces. The weakest point of the paved shoulder and the place where most failures occur is the joint between the travel way and the shoulder. Growth weeds and grass should be treated with herbicides prior to sealing cracks.

Figure (4.5): Maintenance Operation of Bituminous_Shoulders.

Shoulder gutters and curbs are constructed along the outside of the shoulder to collect and channel heavy runoff water into paved ditches or pipes at low points in the grades.

4.3.4 Portland Cement Concrete

Shoulders made of Portland cement concrete should receive the same maintenance as prescribed for rigid pavement.

Figure (4.6): Portland_Shoulders Maintenance Operation.

4.4 Maintenance of Roadsides and Drainage

Roadsides are the areas between the outside edges of the shoulders and right-of-way boundaries. Unpaved median area between inside shoulders of divided highways and areas within interchanges are included. Roadsides should be maintained as nearly as practicable in a manner that contributes to safety, appearance, convenience and pleasure of the public and the preservation and protection of the roadway.

The largest budget allocations for roadside maintenance will be spent for vegetation control on the many acres of roadside right-of-way containing field grass, brushes, and native trees.

Figure (4.7): Maintenance Operation of Roadsides.

Highways are designed with drainage systems to keep water from damaging the structure by controlling or directing the free flow of water over, under or adjacent to the roadway. The drainage system of a modern highway should be designed by an engineer who has a knowledge of hydraulics and highway drainage practices. Many factors should be taken into account during the design stage including: the amount and rate of the rainfall; natural drainage features; water table; contour of the land; effects of hydraulic pressure; etc.

Figure (4.8): Drainage Maintenance Operation.

Maintenance functions include:

1. Keeping watercourses free from accumulations of dirt, debris, vegetation and other obstructions.

2. Correcting malfunctioning parts of system, e.g. erosion and breaks or shifts in structures, waterways, conduits and paving.

3. Anticipating problems and making limited changes or modifications.

4.5 Maintenance of Bridges pavement

Maintenance of bridges is affected by the bridge design, location, age, and use. Bridges are critical to the highway operation. Bridges represent about 30% of highway costs.

Prior to maintaining of pavements of bridges, care should be taken about the reinforcement of deck. The exposed steelwork should be cleaned using sandblasting or other means.

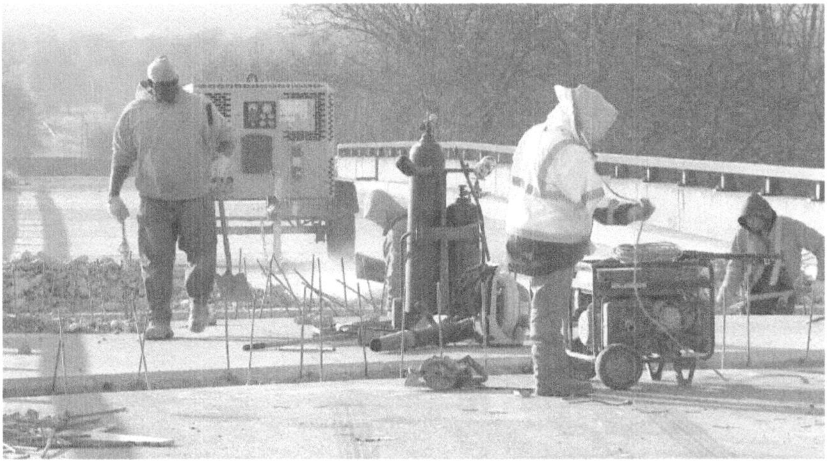

Figure (4.9): Bridge Maintenance Operation.

Deck joints may extrude or become filled with dirt so cleaning and resealing are necessary. On occasion, vehicles out of control strike handrails or other appurtenances and these also must be repaired. If bridge decks become rough or slick, resurfacing should be considered. In this instance surfacing weight may be a primary control on what is used. Remedial measures are sometimes required to correct serious scour around and under piers and abutments.

Because bridge maintenance requires special skills, it is common practice among highway agencies to have travelling crews exclusively for bridgework. Painting and other specialty work is often carried out under contract.

Deterioration of concrete bridge decks under the combination of freezing-thawing and deicing salts has become a major maintenance item especially in cold weather countries. Commonly, the problem begins when the salt penetrates to and corrodes the reinforcing, resulting in spalling of the overlay concrete. Correction may require removal of the concrete, cleaning the steel, and applying new material, possibly polymer concrete.

Sometimes, sealants or overlays of asphaltic materials are employed to provide further protection, such major repairs do not cost much but may cause serious delays and inconvenience to population.

Figure (4.10): Bridge Maintenance Operation.

4.6 Traffic Surface Maintenance

Traffic surface includes such continuing functions as stripping, sign repair and maintenance of street and highway lights and signals. Generally, these are performed by special cross of the highway agency although streetlight and even traffic signals maintenance may be turned over to the local utility company.

Figure (4.11): Operation of Traffic Surface Maintenance.

The signal system or traffic control devices can be damaged which may result in traffic accidents or traffic jam. To keep such systems operating a highly skilled personnel shall be called for the maintenance work. Traffic service, also, include coping with emergencies. For example, during heavy storms maintenance personnel will be on patrol to keep the roads open, sign or barricade washes out and rescue stranded materials.

4.7 Snow and Ice Control

Snow removal is the major winter maintenance problem in roads and highways. Snow removal shall immediately start after snowing according to preplans prepared by the city.

Figure (4.12): Bridge Maintenance Operation.

4.8 Pavement Maintenance Equipment

There are a lot of machines and equipment used in paving works c in the execution of highway projects. In this section of the book wi be reviewed the most important machines and equipment used in th maintenance of pavement as shown in the following pictures:

ASPHALT PAVERS

TANDEM VIBRATORY ROLLERS

SKID STEER LOADERS

ROAD RECLAIMERS

PNEUMATIC ROLLERS

COLD PLANERS

VIBRATORY SOIL COMPACTORS

MOTOR GRADERS

CHAPTER FIVE

5 PAVEMENT MAINTENANCE MANAGEMENT

5.1 Introduction

Historically, only two principles that agencies considered when deciding which roads would be preserved or rehabilitated. These were "Worst First" which gives the first attention to the roadways that looked bad, and "Political Priorities" which are motivated by citizen requests. The current pavement maintenance practice applied to provide the Maintenance and Rehabilitation (M&R) treatments based on subjective judgment, engineering experience or individual opinion. The causes for the ineffectiveness of different maintenance treatments often remain unevaluated because of no availability of objective data base needed for the analysis of various processes, such as planning, design, construction, maintenance and rehabilitation, etc.

5.2 Pavement Management

Pavement management is a systematic process for maintaining, upgrading, and operating physical pavement assets in a cost-effective manner. The process combines applications of established engineering principles with sound business practices and economic theories, thus assuring an organized and scientific approach to decision making. The process involves the following steps for a given pavement section:

1) Assess present pavement condition,

2) Predict future condition,

3) Conduct an alternatives analysis, and

4) Select an appropriate rehabilitation strategy.

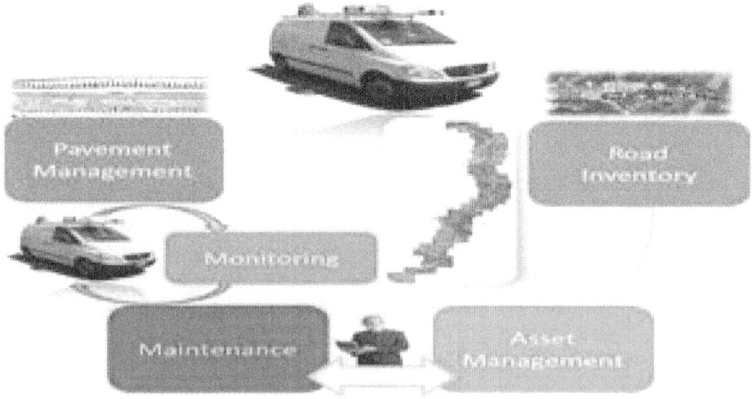

Figure (5.1): Pavement Management Operation.

The pavement management process is strategic and seeks answers to questions such as:

1) What is the current condition of the pavement network?
2) What resources in time, material and personnel will be required to maintain the network at a specified performance standard?
3) What should be the annual work program to address the most critical needs that reflect available resources?

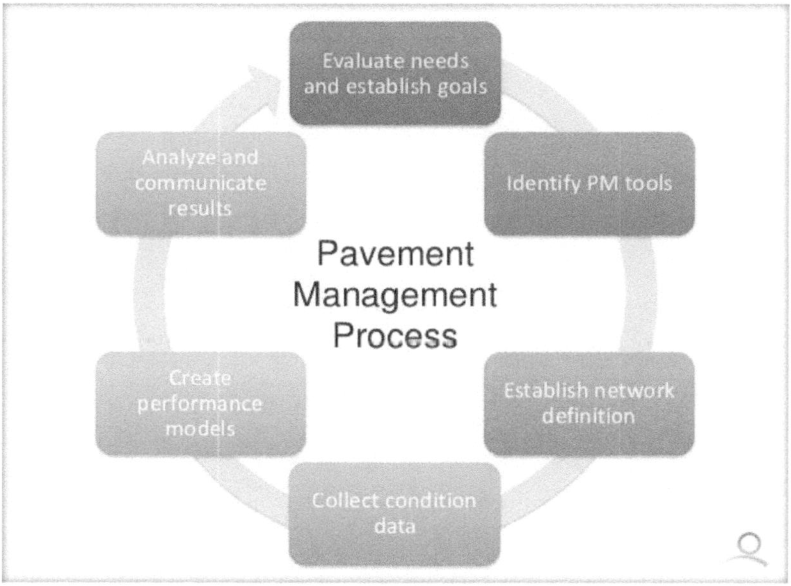

Figure (5.2): Pavement Management Process.

5.3 Pavement Management Plan

Success in pavement management is accomplished by maintaining pavements in good or better condition while maximizing their life cycle. This can be achieved by monitoring a pavement's condition over time by means of a data collection and rating procedure combined with effective utilization of a pavement management system to optimize strategic application of pavement preservation treatments.

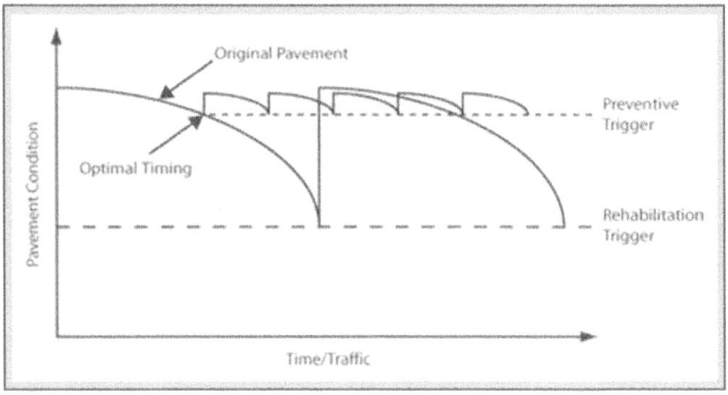

Figure (5.3): Pavement Management Plan.

A pavement management plan involves adopting a strategy of activities that monitor pavement conditions and prioritize maintenance, preservation and construction activities with available funding to optimize the conditions of a roadway network.

An effective pavement management plan assists transportation agencies to optimize their available funding, better communicate funding needs, and more objectively manage the overall pavement network. Pavement management plan key components

1) System condition assessment and performance monitoring (pavement data collection, analysis and review)

2) Effective pavement preservation practices

3) Agency goals and policies

4) Public education and engagement

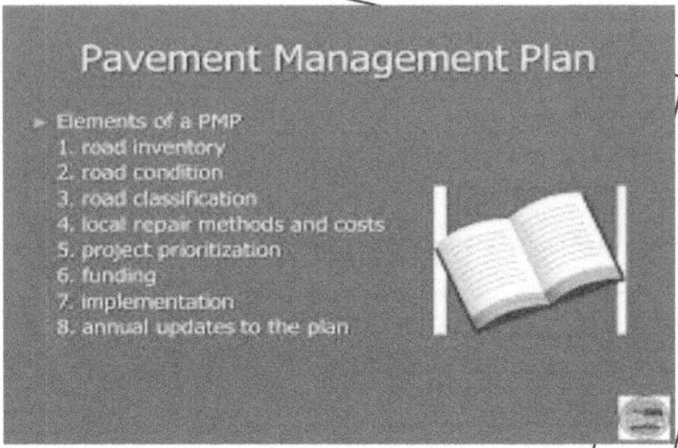
Pavement Management Plan
Elements of a PMP
1. road inventory
2. road condition
3. road classification
4. local repair methods and costs
5. project prioritization
6. funding
7. implementation
8. annual updates to the plan

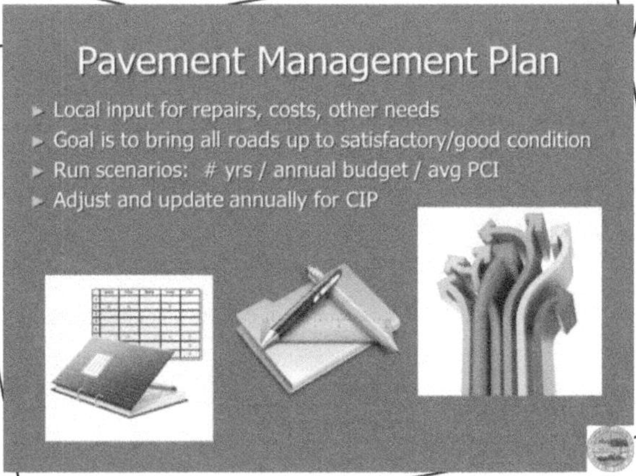
Pavement Management Plan
Local input for repairs, costs, other needs
Goal is to bring all roads up to satisfactory/good condition
Run scenarios: # yrs / annual budget / avg PCI
Adjust and update annually for CIP

5.4 Level of Pavement Management

Pavement management can occur at the project level and the network level as illustrated in Figure (5-4). Network level management relates to the evaluation of all pavements under an agency's responsibility. The main goal of network level management is to prepare an agency-wide prioritized pavement repair program that will yield the greatest benefit or least total cost under overall budget constraints. Also, network level management depends on more approximate data than project level management. Project level management is based on a particular location and it usually comes after the analysis of network level in local agencies. Once a section has been identified for repair at the network level, then an engineering analysis is performed at the project level. A more detailed evaluation is required through this analysis, since the information collected at the network level does not normally contain the type of data required to make detailed decisions for an individual project design. During a project level analysis, additional testing, such as nondestructive and coring testing, is often conducted to give additional information about pavement condition and the cause of deterioration.

In general, network level analysis is of more interested to budget irectors and senior decision makers and doubtless it is the most owerful of pavement management properties because it involves:

1) Identification and ranking of pavements needed maintenance.
2) Budgeting of network-level.
3) Forecasting of long range budget.
4) Assessment of network-level pavement condition.
5) Forecasting of future pavement conditions.
6) Selecting and designing the desired solutions.

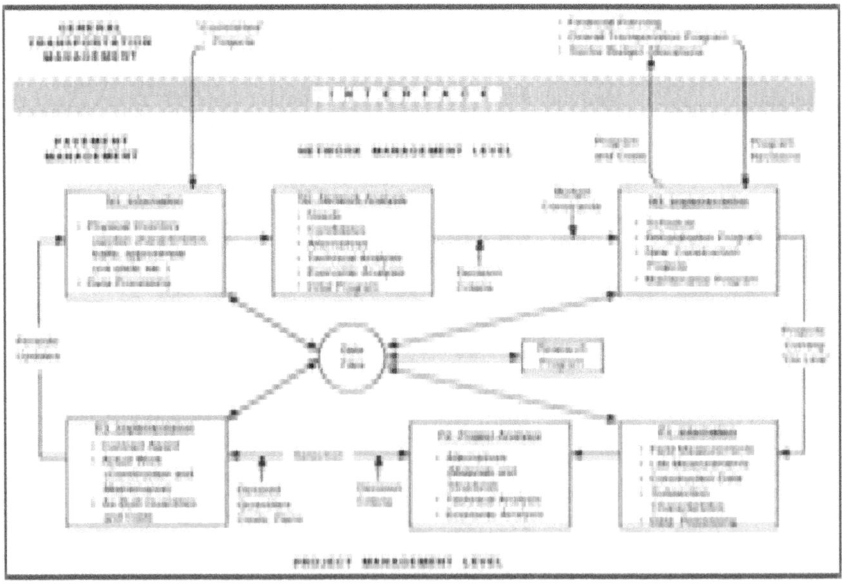

Figure (5-4): Levels of Pavement Management.

The project level analysis is of greater interest to the technica staff. It involves:

1) Evaluating the causes of deterioration.
2) Identifying potential solutions.
3) Evaluating benefits of the alternative by life-cycle costing.

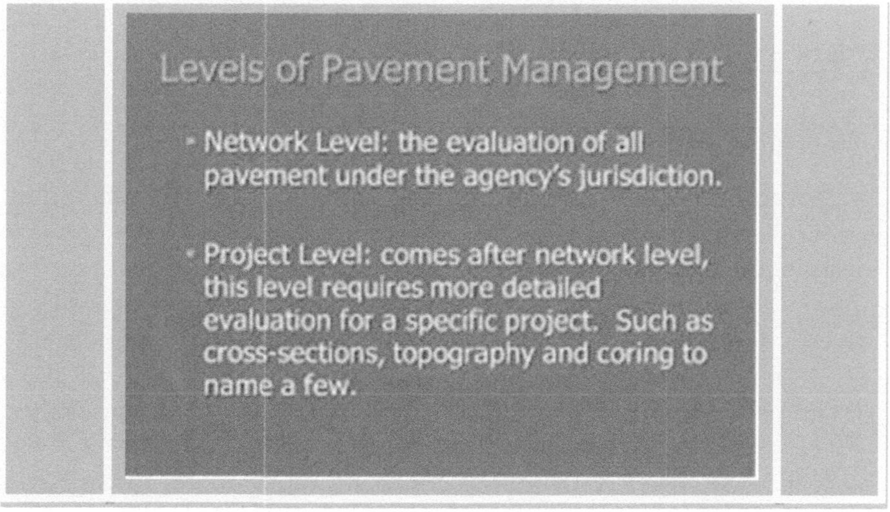

5.5 Types of Pavements

As previously mentioned, there are two main types of pavements: rigid and flexible pavements.

.5.1 Rigid Pavements

Rigid pavements consist of a sub-grade/sub-base foundation overlaid by a concrete slab. The slab should have an enough depth to avoid possible premature failure owing to traffic loading. Failure of pavement can also be generated as a result of other causes as will be explained later, therefore appropriate measures should also be taken into account. It is understood that the concrete mix proportions provide clear idea about both strength and the resistance to factors such as climate changes. The slab thickness is of very important factor and it can be determined by the traditional design procedure. Joints in the concrete are required to enhance the resistance of concrete to tensile and compressive forces owing to shrinkage effects. The concrete of pavement can be either plain (non-reinforced) or reinforced with steel bars. The following Figure clarifies layers of the rigid pavement.

Rigid highway pavements can be divided, based on surface material, into three general types: plain concrete pavements, simply reinforced concrete pavements, and continuously reinforced concrete pavements. The term of each pavement type is related to the amount of reinforcement used. The description of the three types are listed below:

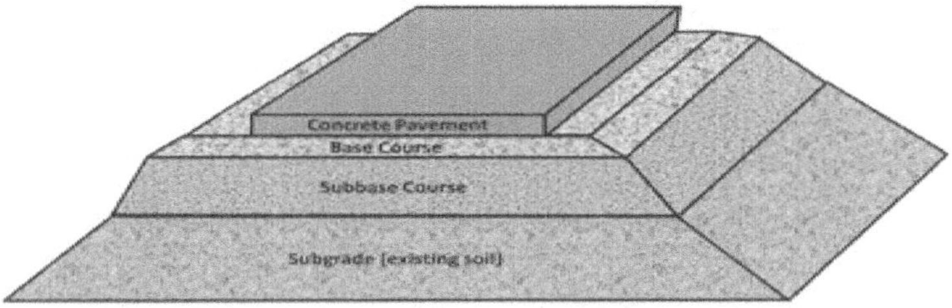

Figure (5-5): Components of Rigid Pavements.

5.5.1.1 Jointed Plain Concrete Pavement

Plain concrete pavement has no steel or dowels for load transfer. However, steel tie bars often are used to provide a hinge effect at longitudinal joints and to prevent the opening of these joints. Plain concrete pavements are used mainly on low-volume highways or when cement-stabilized soils are used as sub-base.

Joints are placed at relatively shorter distances (10 to 20 ft) than with other types of concrete pavements to reduce the amount of cracking. So it is necessary to know the joints types and the description of each type.

To allow for movement produced by changes in temperature, joints shall be included in the pavement slab design . The shrinkage and temperature stresses are released by transverse joints which located across the pavement at right angles to the center line. The biggest effect of the stresses is in the longitudinal direction and therefore the longitudinal joints manage those induced stresses most evident across the width of the pavement. four main types of transverse joints can be observed in pavements:

1) Contraction joints: These kind of joints release stresses by permitting the adjacent slab to contract, thus reduce in tensile stresses within the slab. The contraction joints, should be opened to allow the movement and at the same time eliminating the vertical movement that may occur between adjacent concrete slabs.

2) Expansion joints: These joints create full discontinuity between the two adjacent sides. They are usually filled with a compressible materials to allow the adjacent concrete slabs to expand. These joints can also act as warping or contraction joints.

3) Warping joints: These joints are usually used in plain (unreinforced) concrete slabs only, particularly, in long narrow concrete lanes where warping stresses may occur. These joints allow small angular movements to take place between adjacent concrete slabs.

4) Construction joint: Construction joints do not allow for any relative movement. These kind of joints are used when the construction of slab ends for any reason before reaching the intended contraction or expansion joints as planned.

5.5.1.2 Simply Reinforced Concrete Pavement

Simply reinforced concrete pavements usually have longitudinal dowels for the purpose of transferring the traffic loads across joints. These joints are usually located at distances between 30 and 100 ft. Temperature steel is also used within the concrete slab, with an amount depending on the length of the slab, while tie bars are also used in longitudinal joints.

5.5.1.3 Continuously Reinforced Concrete Pavement

The presence of reinforcement in the continuously reinforced concrete pavements cancels the need for transverse joints. Construction joints or expansion joints are used in the reinforced concrete slab when necessary, such as at bridges. These kind of pavements have a relatively high percentage of reinforcement, usually at 0.6% of the cross section of the slab. Tie bars across the longitudinal joints are also used.

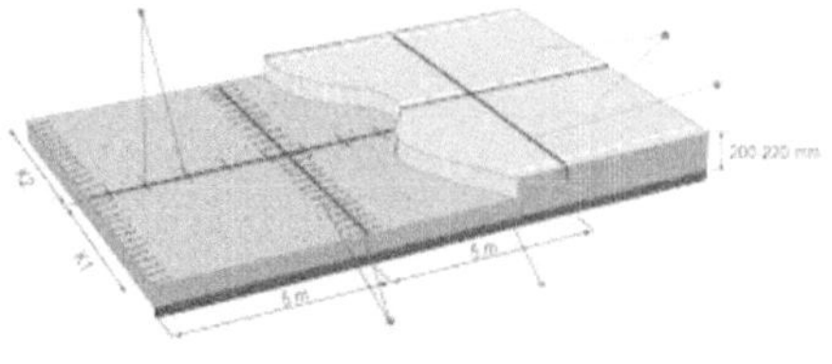

5.5.2 Flexible Pavements

The surfacing and road-base materials, bound with bitumen binder, overlay granular unbound or cement-bound material. As it is known, the flexible pavement consists of many layers including:

1) The sub-grade layer: is usually the natural material located along the horizontal alignment of the pavement and serves as the foundation of the pavement structure. It also may consist of a layer of selected borrowed materials, well compacted to meet the specifications. It may be necessary to treat the sub-grade material to achieve certain strength properties required for the type of pavement being constructed.
2) The sub-base layer: this layer is located immediately above the sub-grade. The sub-base component consists of materials of a superior quality comparable to that used for sub-grade construction. The requirements for sub-base materials usually are given in terms of the gradation, plastic characteristics, and strength. In cases where suitable sub-base material is not readily available, the available material can be treated with other supplements to achieve the necessary properties. This process of treating soil to improve its engineering properties is known as stabilization.

3) The base layer: The base course lies immediately above the sub-base layer. Sometimes, it is placed immediately above the sub-grade if the sub-base course is not used. The base course usually consists of granular materials such as crushed stone, granulated or grinded slag, crushed or uncrushed gravel, and sand. The specifications for base course materials usually include more strict requirements than those for sub-base materials, particularly with respect to their plasticity, gradation, and strength.

4) The wearing surface layer: it is the upper course of the road pavement and constructed immediately above the base course. The surface course in flexible pavements usually consists of a mixture of mineral aggregates and asphalt. It should be capable of withstanding high tire pressures, resisting abrasive forces due to traffic, providing a skid-resistant driving surface, and preventing the penetration of surface water into the underlying layers. The thickness of the wearing surface can vary from 3 in. to more than 6 in., depending on the expected traffic on the pavement.

5) 5.5.3 Composite Pavement

They are rigid paving roads to which a layer or more of asphalt concrete is added. A composite pavement is a type of pavement that utilizes both asphalt and concrete. ... This pavement type can also be used in conjunction with Roller-Compacted Concrete (RCC) pavements, where the RCC pavement provides the structural capacity that the conventional concrete pavement base would.

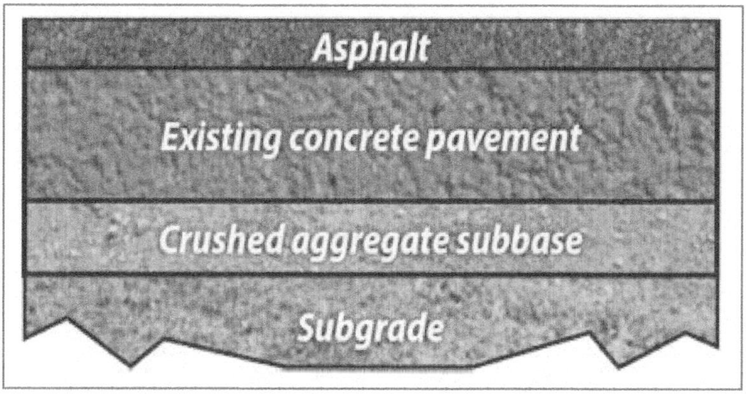

Figure (5-6): Components of Composite Pavements.

Figure (5-7): Components of Flexible Pavements.

The difference between Flexible Pavements and Rigid Pavements can be shows in Table (5.1) and Figure (5.8).

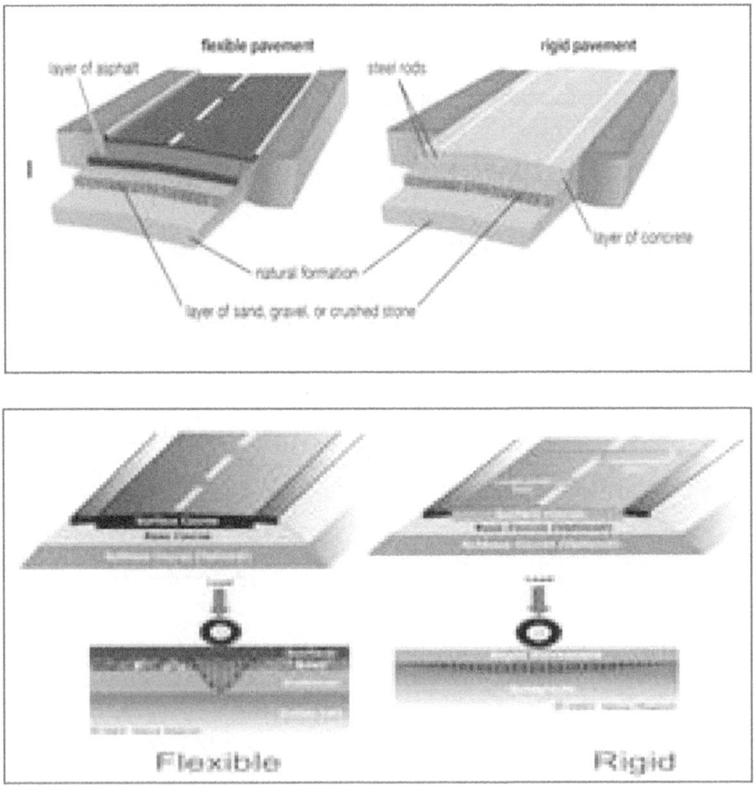

Figure (5-8): Difference Between Types of Pavements.

Table (5.1) Difference Between Types of Pavements

Flexible Pavement	Rigid Pavement
It consists of a series of layers with the highest quality materials at or near the surface of pavement.	It consists of one layer Portland cement concrete slab or relatively high flexural strength.
It reflects the deformations of subgrade and subsequent layers on the surface.	It is able to bridge over localized failures and area of inadequate support.
Its stability depends upon the aggregate interlock, particle friction and cohesion.	Its structural strength is provided by the pavement slab itself by its beam action.
Pavement design is greatly influenced by the subgrade strength.	Flexural strength of concrete is a major factor for design.
It functions by a way of load distribution through the component layers	It distributes load over a wide area of subgrade because of its rigidity and high modulus of elasticity.
Temperature variations due to change in atmospheric conditions do not produce stresses in flexible pavements.	Temperature changes induce heavy stresses in rigid pavements.
Flexible pavements have self-healing properties due to heavier wheel loads are recoverable due to some extent.	Any excessive deformations occurring due to heavier wheel loads are not recoverable, i.e. settlements are permanent.

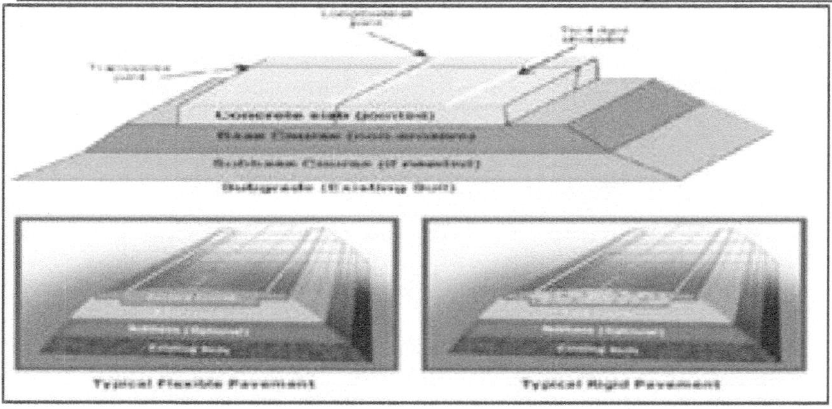

5.6 Highway Pavement Material

There are many types of raw materials used in highway projects., Each material will be briefly discussed in the following next sections.

5.6.1 Soils at Sub-Formation Level

Unless the in-situ soil is composed of hard soil or rock, it is likely to be week to carry loads of construction and traffic. Thus, it is required to construct additional layers of materials to resist the stresses transferred to the subsoil. In other words, the natural soil in site would permanently deformed if it is loaded by stresses generated by heavy loading of vehicles. In general, the shear strength and modulus of stiffness are accepted indicators to determine the resistance of the soil to permanent deformation. Soils having high values of stiffness modulus and shear strength will be less susceptible to permanent deformation and vice versa. The two properties of soil are reduced owing to the increase in moisture content.

Determining the two properties is very essential to define the suitable thickness of pavement layers. If it is not available to determine the two properties (shear strength and stiffness modulus), the California Bearing Ratio (CBR) test is usually used as an indicator. While it is not a direct measure of either the shear strength or the modulus of stiffness, the CBR test is a widely acceptable and used.

The CBR test is an attempt to mimic the behavior of soil when subjected to forces such as the load of wheels. Although it was developed in California before World War II, the CBR test still the basis for the empirical pavement design methodology everywhere. It should be noted that the test does not determine any fundamental strength characteristic of soil. It mainly involves a cylindrical plunger being driven into a soil at a standard rate of penetration. The level of resistance of the soil to the penetrative effort is recorded.

The test can be performed either on site or in the laboratory. If the test is carried out in the laboratory, it is required that the dry density and moisture content of the sample being tested as closely as possible to the expected values of the in-situ soil. Particles having diameter greater than 20 mm should be removed. If the CBR test is carried out in-situ, it should be performed on a newly exposed soil surface at a depth that seasonal variations in moisture content would not be reached.

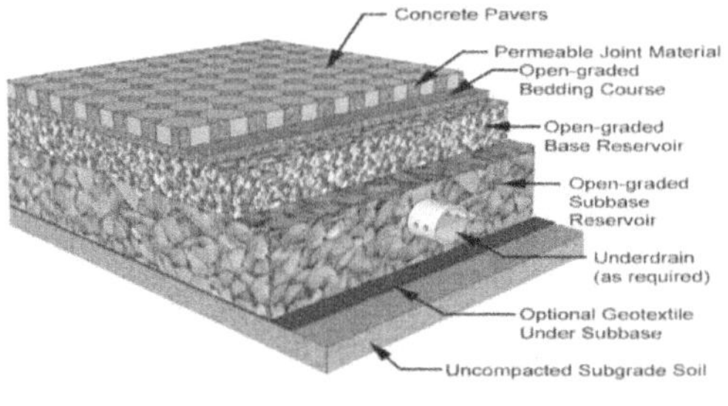

5.6.2 Sub-Base

Both the sub-base and capping work as a regulator and protector of the surface of the sub-grade located below the sub-base from influences of severe weather. The sub-base along with the sub-grade, represent the secure platform on which the upper layers of the highway pavement can be constructed.

The thickness of sub-base of the pavement relies on the strength of the sub-grade, while it independent of the cumulative traffic incident on the upper layers of the pavement over its service life. The thickness of the sub-base decreases as the CBR of subgrade increases. For example, the required sub-base depth is approximately 225 mm for sub-grades in excess of 5% CBR. This thickness decreases to about 150 mm at a sub-grade CBR of 15%. Granular and treated soil sub-bases are suggested for flexible pavements while cemented sub-bases may be enough for rigid-type pavements.

The following five broad approaches should be taken into account when the sub-base layer is applied:

1) No sub-base layer is required if the sub-grade layer is consisted of hard rock or of a granular material with a CBR value of at least 30%, providing that the water table is not at a high level.

2) In the case of sub-grades layer with a CBR greater than 15%, a sub-base thickness of 150 mm is necessary (in practical terms this constitutes the minimum sub-base layer thickness for ensuring proper spreading and compaction).

3) A layer of 150 mm thick of sub-base on top of capping would be required for pavements having sub-grade' CBR is lower than 2.5% and for construction of rigid pavement having materials with CBR lower than 15%. The thickness of the capping layer can reach as much as 600 mm thickness where the CBR of the sub-grade is below 2%.

4) Soil replacement should be considered when the sub-grade CBR value is less than 2%. More suitable materials with CBR higher than 5% should be used. The thickness of the imported materials may be ranged from 500 mm and 1000 mm deep. The imported soil usually assumed to have a CBR of 2% and thus may need about a 600 mm capping layer.

5) Where the CBR value of the sub-grade lies between 2.5% and 15%, two options are available:

Use 150 mm of sub-base layer over a layer of capping material having a thickness depending on the CBR of the sub-grade layer, or

When the layer of sub-base have CBR value greater than15% the layer thickness should be 150 mm at least and 350 mm (at 2.5% CBR) in thickness.

5.6.3 Materials within Flexible Pavements

Materials used in flexible pavements may include the following.

5.6.3.1 Bitumen

Bitumen is one of the products of refining process of crude oil. It represents the basic component of the surface layers in pavement construction. In addition to its ability to resist both deformation and changes in temperature, the binding effect of bitumen reduces the loss of materials from the surface of the pavement and inhibits the penetrating of water in the pavement structure. Bitumen can be found in two forms:

Tar – which is resulted from the production of coal gas or the coal production

Bitumen – which is a product of the oil refining process. Considering the relatively low production of tar, bitumen is the most commonly used as binding and water resisting material for pavements.

5.6.3.2 Asphalts

Two types of asphalts will be discussed in this section:

1) Hot rolled asphalt: It is a form of asphalt that usually consists of fines, binder, filler and mixture of aggregates. The hot rolled asphalt has low air void content and dense texture. The grading normally contains a higher proportion of both fines and binder compared to other components. The content of coarse aggregate is typically ranged from zero to 55%. This content is changed based on the type of layer of asphalt..
2) Mastic asphalt: is water proof and durable heavy-duty material that usually used as wearing course material. Components of the mastic asphalt are asphaltic cement of low-penetration grade bitumen, fine aggregate and filler. The proportions of components result in a low-void impermeable mass. The mastic asphalt contains sometimes coarse aggregate, which must pass sieve of 14 mm openings and retain on sieve of 10 mm openings. The content of the fine aggregate (particle size between 2.36 mm to 0.075 mm) should be between 45% and 55% . The mastic asphalt must also contain high proportions of both binder and filler material.

5.6.3.3 Aggregates

Aggregates are used in the pavement construction as a filler and to enhance the strength of the pavement. The thickness of the pavement layer and the requirement of the surface texture identify the maximum nominal aggregate size. The thickness of the pavement layers should generally higher than 1.5 times the maximum aggregate size. Large size of aggregate should be avoided in the pavement as these sizes can be possibly crushed under the weighs determined from both the required thickness and the surface texture of the pavement. The size of aggregate must not be greater t of rollers and compactors.

5.6.4 Materials of Rigid Pavements

Reinforced and non-reinforced concrete are the main materials used in rigid pavement .

5.6.4.1 Slab Concrete

Specifications of concrete mixture used in the pavement construction can be changed from one country to another. Generally, ordinary Portland Cement (ASTM Type I) is used in pavement concrete. However, sulfate resistance cement (ASTM Type V) can be used is sulfate attack is expected from surrounding soil or water. Most of specifications limit the minimum cement content by 320 kg/m^3 and maximum water/cement content by 0.5.

The required compressive strength of the pavement concrete is determined by the designer. Because of the internal stresses created by climate, the use of joints is very necessary to reduce these stresses. The pavement is divided into series of slabs by joints to permit the release of induced stresses, which can be more destructive than the stresses created by traffic loads. The use of reinforcement can also reduce the effect of stresses induced by climate. Reinforced slab with movement joints is called Jointed Reinforced Concrete (JRC). However, type of joints and location are important to be determined in the design of concrete pavement.

5.6.4.2 Reinforcement

Prefabricated mesh or a bar-mat can be used as reinforcement in concrete of pavement. In addition to its contribution to the strength of concrete, the reinforcement limits the extent of surface cracking and lessens the effect of temperature stresses. The reinforcement should be thoroughly cleaned to ensure good mechanical bond with concrete. Reinforcement that decreases the surface cracking should be close to the top surface of concrete. The concrete cover should be determined based on thickness of the slab, diameter of reinforcement and the maximum size of aggregate. In general, the reinforcement should be stopped approximately 125 mm from the edge of concrete slab, 300 mm from any transverse joint and about 100 from longitudinal joints. It is worthy stating that different codes and building codes may have different numbers and conditions for the use of reinforcement in concrete of pavement.

5.7 Principle of Pavement Maintenance

Maintenance can be, generally, defined as a mixture of any processes taken to keep certain material in service, or to get it back to its normal condition. Subsequently, maintenance keeps materials in good conditions and in service, so as to do their job as required.

Highway maintenance, is a program to preserve, and restore a system of roadways with its elements to its designed configuration [14]. Road maintenance can be defined as a way for keeping the road always in good condition, as when it was first built. Moreover, maintenance provides safety, and the appropriate and economical movement from one place to another.

Maintenance can also be defined as a way of getting the road back to the condition when it has been first built, or it can be defined as periodic activities working under the normal circumstances of environment and traffic, to keep the pavement in a condition almost similar to that when it was first constructed. Maintenance also aims at keeping the road in service for as longer period as possible.

In addition to ensure the passing of vehicles safely in the road. Moreover, maintenance ensures a good investment of money spent in construction the road. Subsequently, highway maintenance must be focused in two essential sides as follows:

First: - Maintenance of highway structure by keeping all road structural elements such as (asphalt layers, base and sub base layers, and shoulders) in good condition almost similar to that when the roads were first constructed.

Second: - Maintenance of road services, including traffic signs, fences, wire fences, phones illumination poles, roadside, etc. in such a way ensures a safe and economical

Since the maintenance activities are different from one location to another according to their circumstances and abilities, the following definitions are identified to indicate these activities as follows:

(1) Preventive maintenance: -

A preventive maintenance can be defined as a systematic approach to apply number of preventive maintenance treatments over time. By applying these treatments, not only the quality of the pavement surface will be improved and the pavement life will be extended, but also the true benefits are realized when these maintenance treatments are performed at a consistent schedule. An effective pavement preservation system integrates many preventive maintenance and rehabilitation treatments. The main goal of such a system is to enhance system-wide performance and extend pavement life in a cost-effective and efficient method. Studies stated that applying preventive maintenance for six to ten times is more cost-effective than a "do nothing" strategy. Critical elements of the successful pavement maintenance program are:

1) Choosing the roadway

2) Defining the cause of the problem

3) Determining and applying the correct treatment(s)

4) Defining the correct time to do the required work

5) Observing performance

Preventive maintenance works can include conventional maintenance strategies such as crack sealing, chip sealing, thin overlays, fog sealing, and rut filling. They can also include emerging treatments such as very thin overlays, ultra-thin wearing courses, and micro surfacing applications.

It is a process of finding out the damages and repairing them when they first appear to prevent the damages before they become serious. The passing vehicles with loads, particularly those exceeding the designated loads, lead together with the bad weather conditions to increase and spread the damages within few days. The matter, which requires implementing detailed and continuous periodic examinations, is the walking of specialists on foot on paving or using highly observation devices.

2) Corrective Maintenance

Corrective maintenance activities differ from preventive maintenance activities primarily in cost and timing. Corrective maintenance activities are performed when there is urgent need to repair the pavement, and is therefore more costly, unlike preventive maintenance which is performed when the pavement is in good condition.

Corrective maintenance works are more reactive than preventive maintenance, and must be performed to correct a specific section of pavement or area of distress. Also, any delays in maintenance may increase pavement defects and their severity and when the distress is corrected, the cost is much greater. Consequently, when corrective maintenance is performed, the life cycle costs of pavement will be considerably increased. The works of corrective maintenance include structural overlays, mill and overlays, crack repair, pothole repair, and patching.

(3) Rehabilitations:

They are activities through which the roads are restored to their original state after they disintegrated and damaged where the periodic maintenance fails to repair them. Subsequently, these activities cause a substantive change in the paving constructive condition as adding asphalt layers and reconstruction. Primary objectives of pavement rehabilitation, as indicated in the ASSHTO policy are to:

a) Improve surface smoothness;

b) Extend pavement life;

c) Improve skid resistance;

d) Reconstruct sections having poor foundations; and

e) Improve drainage.

(4) Improvements:

These are activities of road parts reconstruction such as enlarging the horizontal curves, adding supported walls or paving shoulders in order to increase road safety, and raise the service state. However, these activities do not increase the paving endurance force.

5) Upgrading:

Upgrading are activities aim at increasing the paving endurance ɔrce. These may be accompanied by an increase in road's capacity ɔath number) in order to contain the new traffic volume and it's xial loads.

5.8 Pavement Maintenance Classification

Maintenance can be classified according to either the activities recurrent periods included there in or according to the activity type or their aim as shown in the following diagram:

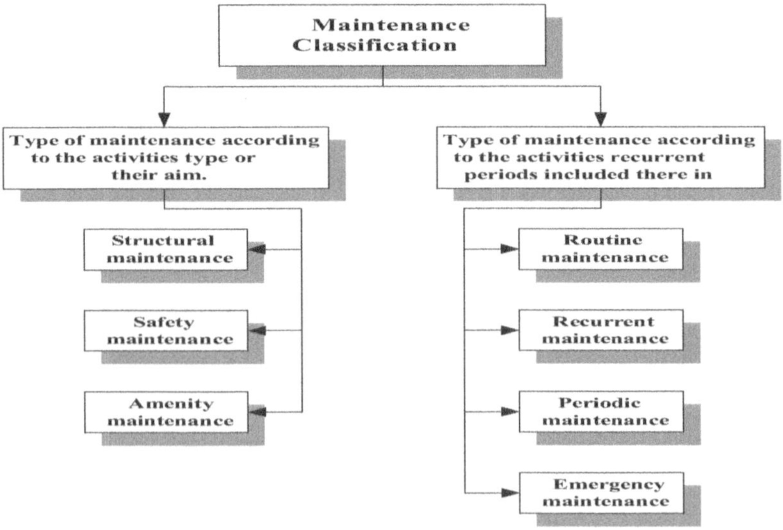

Figure (5.9): Maintenance Classification.

5.8.1 Routine Maintenance:

It is a group of activities that aims at keeping the road safe through daily examination by the road control officials on bot sides. The activities may include some immediate and protecti measures, including the removal of all obstacles and the cleanir the road surface, the mediate, shoulders, and sides road. Generall the absence of the maintenance leads the road to more collapse subsequently, the road loses a lot of its technical specifications an as a result, it will be useless.

From the economical viewpoint, routine maintenance must be interested to prolong the road's life and avoid the necessity for recurrent maintenance or reconstruction.

5.8.2 Recurrent Maintenance

It is a group of activities that must be implemented on consecutive periods. This kind of maintenance depends on the traffic volume. An example of recurrent maintenance is repairing the crack

5.8.3 Periodic Maintenance

It is a number of activities that must be performed in intervals to the extent of certain number of years, in order to repair the big damages that may occur in some parts of the road due to usage for a long period of time. Implementing such maintenance requires necessary advanced preparations.

An example of such maintenance is the addition of new asphalt layers or removing the damaged parts and reconstructing after handling the reasons of damages or changing the pointed curves, or widening some roads.

·Generally, the absence of the routine maintenance will lead to the implementation of periodic maintenance, which costs huge sums of money.

5.8.4 Urgent or Emergency Maintenance

Emergency maintenance is performed during an emergency case, such as when a severe pothole must immediately be repaired, generally, for permitting for traffic to use the roadway or for safety reasons.

Some of the typical considerations for selecting a treatment strategy are no longer important when emergency maintenance is needed. Cost may be the least important consideration than safety and time of maintenance application. Materials that may not be acceptable for the purpose of implementing preventive or corrective maintenance works for cost or long-term performance causes may be highly acceptable in an emergency maintenance situation.

Emergency maintenance is the kind of maintenance that must be implemented immediately and without any delay to repair any damage that may occur in the roads networks. Such as damages resulted from traffic accidents or the blocking of the road by sand hills or floods, etc.

Any delay in implementing such maintenance will cause loss of money, therefore, most of the countries pay great attention to provide well-trained and well-equipped urgent maintenance teams.

5.8.5 Structural Maintenance

It is the group of activities that aims of maintaining the road structure including the paving surface, sewage and etc..

5.8.6 Safety Maintenance

It is the group of activities that aims of maintaining parts of the road including illumination, safety fences, wire fences, traffic signs, etc. to ensure the safety of people using the road.

5.8.7 Amenity

It is a group of activities that provide comfortable circumstances for people using the road such as taking care of trees and the areas overlooking the road.

5.9 Importance of Pavement Maintenance

First: **Reducing highway deterioration.**

Figure (5.9), which is a typical performance curve, clarifies the relation between the highway condition and its age. When the highway gets older due to its use by vehicles, gradual deterioration appears. Generally, the deterioration rate depends on many factors including weather condition, paving quality, underlying soil strength, traffic volume, and axial loads for the heavy vehicles allowed to use the highway.

Neglecting routine maintenance, such as neglecting the small cracks on the paving surface, will cause gaps through which water penetrates into base layer or the layer below the base one. This matter leads to sink in certain parts of the highways. Subsequently, repeated periodic maintenance is required whose cost are (10) times more than the routine maintenance cost .

From what has been mentioned above, it can be recognized the economic advantages of implementing maintenance works properly. It can be concluded, that the highways after a certain period of time require maintenance works. Improvements and rehabilitation of highways are important to keep them as a longer period as possible in service as shown in Figure (5.10), which indicates a new circle of the highway's life.

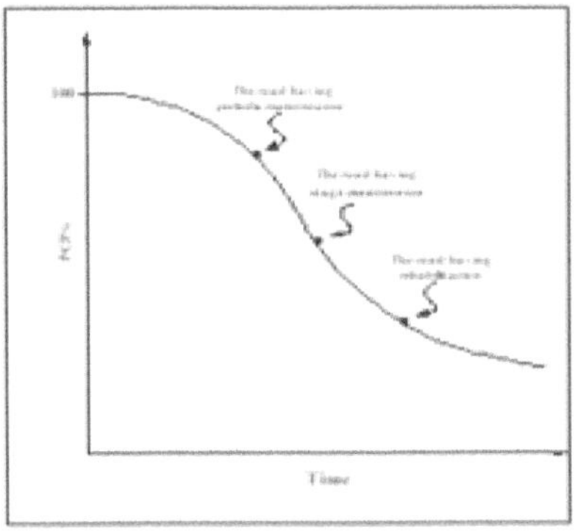

Figure (5.9) Performance Curve

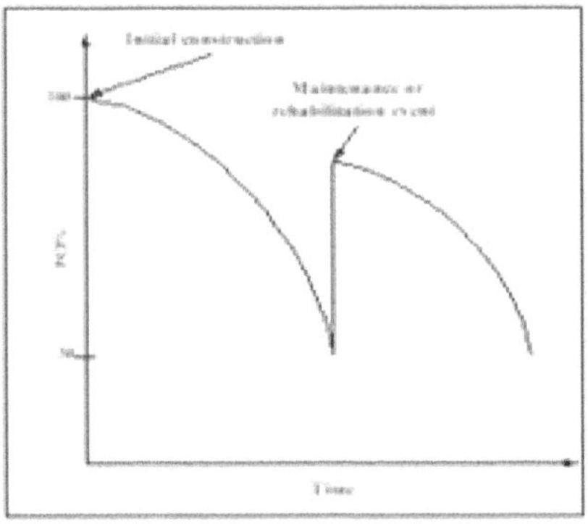

Figure (5.10) pavement performance

***Second:* Lowering Vehicle Operating Cost.**

Driving vehicles in damaged and narrow highways will increase the operating cost, because of more fuel consumption. Statistical studies, in this respect, were published lately in the United States of America indicate that the waste of fuel is about (16.4) billion gallons per year because of using damaged highways.

In addition, damaged highways will increase the vehicles wheels consumption, as well as, the suspension system is consumed in a shorter time.

Subsequently, everyone in the United State of America has to pay a hidden tax which is about (136) dollars every year.

Thus, keeping highways new by maintenance will provide a comfortable and easy driving and reduce the cost borne by highway users. Subsequently, a great benefit for the national economy will be achieved by paying attention to maintenance and improving the highway to lower the vehicles operating cost .

Third: **Keeping the Highway Open.**

The third justification for implementing maintenance works is to keep the highway open to ensure providing continuous service for residential, industrial and agricultural facilities surrounding the highway. Blocking the highway either because of the sand hills or floods in rain seasons will also create series of social and economic consequences. In one of the African countries, 40% of vehicles are unable to reach their destinations due to the bad condition of the highways in rain seasons. The matter, which led to close a number of factories and plants for several months, as well as, delivering their production to distribution centers. Additionally, the agricultural areas at that were affected for not being able to get the fertilizers or delivering their products to the markets.

5.10 Pavement Performance Evaluation

Performance evaluation of pavement is very important in making the decisions of timely rehabilitation and maintenance of pavements on one hand and in setting up of the priority program for maintenance on the other hand. Furthermore, the accumulative information about the performance evaluation of different pavement types and locations can assist designers and decision-makers in prediction of the expected time-cost relations of new highway projects.

The performance evaluation should normally include both the structural stability and the function of the pavement. In structural evaluation, components of the pavement including all pavement layers should be evaluated in relation to their ability to resist the different applied loads. The structural evaluation should clarify if there are deflections and pavement distortions in any of the layers related to stresses and strains generated by traffic or other loads.

The functional evaluation mainly deals with the riding quality of the pavement. In other words, the functional evaluation assesses the serviceability and the general behavior of the pavement to traffic. The functional evaluation should clarify if there are cracks, unevenness, rutting or any other defects that may affect the serviceability of the pavement.

Prior to starting any evaluation of pavement, an inventory of the pavement should be performed. The pavement inventory is collecting of all historical information related to the pavement and information of traffic. The inventory will greatly help the decision-makers in setting up the adequate decision and selecting the suitable maintenance program.

5.11 Pavement Inventory

Hundreds of forms are available in municipalities and highway departments that are used for pavements inventory. Some specifications such as ASTM also offer procedures of pavement inventory. However, the main purpose of the pavement inventory is to assist in design of the suitable maintenance and rehabilitation program through providing accurate information about the proposed pavement. The information of the inventory should include the following:

1) Information about the traffic such as number and type of vehicles
2) Structure of the pavement: This may include date of construction, number of layers, original design of pavement, thickness of layers, and others
3) Identification of pavement: This includes location of the pavement, dimensions of section and accurate information about the pavement such as connected with other pavements and etc.

4) Drainage of pavement: This should include accurate description of the drainage structures of the pavement, their locations, types and others.
5) History of Maintenance: This may be the only part of the pavement inventory that should be timely updated. Details of the previous maintenance carried out on the pavement and its structure should be accurately and clearly recorded. This is of great importance in

It is too difficult to assess the entire pavement because the one network may have various evaluations based on type of distress and or even different traffic loads. Accordingly, the network should be divided into smaller areas to have more accurate evaluation. The evaluation should be established on "Repair Unit". ASTM D5340 suggests dividing the pavement network into

1) Areas that have different uses based on the purpose of use.

2) Each single area to be divided into sections based on pavement history, design, traffic and construction.

3) Each section to be divided into sample units. This can be performed by assuming imaginary joints that divide the one section into sample units.

The inspected units should be accurately identified and also easily founded so the quality control personnel can easily locate them on the surface of the pavement.

5.12 Pavement Condition Index Surveys

There are many test methods used to evaluate the condition of pavement. These methods may focus on one or more features of the pavement. However, presenting of the final results of evaluation as a number "index" is the most understandable and acceptable method among engineers either for pavements or other infrastructure systems. In the index method, numbers should be clearly determined so comparison between different pavements becomes more accurate. This index method is also very beneficial in terms of designing the suitable maintenance program and estimating life-cycle costs of pavements. Some of the most common methods used to evaluate the pavement condition are explained below

5.12.1 Skid Resistance Rating

Skid resistance is very important parameter in evaluation of pavements. It is defined as the force generated when the tire of a vehicle is prevented from rotating slides on the surface of the pavement. Design of pavement should provide users with reasonable safe roadways; otherwise the inadequate skid resistance can result in higher level of accidents. The skid resistance is also indirect measure of material of pavement surface. The skid resistance involves two parameters; the tire of the vehicle and the texture of the pavement, therefore it is very complicated to obtain a clear number of the skid resistance as the two parameters can be significantly changed. It should be noted that the wet wetness of the pavement decreases the skid resistance. It is well known that the skid resistance of the pavement increases during the first two years following the construction as the aggregate of the pavement becomes exposed owing to worn away of the pavement. Then, the skid resistance decreases with time. The skid resistance is also different in the four seasons owing to the wetness of the road. Skid resistance is measured as a friction factor or skid number:

Friction Factor, $f = F/L$ ....(1)

Skid Number (SN) = 100 (f) ....(2)

Where:

F is frictional resistance to motion and L is load perpendicular to interface

The most common methods used to measure the skid resistance are the locked wheel tester, the spin up tester and the surface texture measurement. The following table (5.2) shows the typical skid numbers and their explanations:

Table (5.2) Common Methods to Measure the Skid Resistance

Skid Number	Comments
Less than 30	Take measure to correct
= 30	Acceptable for low volume roads
31-34	Monitor pavement frequently
≥ 35	Acceptable for heavily travelled roads

5.12.2 Pavement Surface Evaluation and Rating

This system was developed by University of Wisconsin-Madison and it basically a scale of 1 to 10. The evaluation of pavement condition in this method is relied on visual inspection. Although results of this method can be used for comparing of different pavement segment, the rating cannot be disaggregated into measurements of distress types. The method can be performed on foot or even by windshield. Pavements can be assessed quickly using this method. The method can be used to assess both flexible and rigid pavements. The rating number can be translated into condition assessment as shown in the table (5.3) below:

5.12.3 Pavement Condition Index (PCI)

One of the most widely accepted index in the condition assessment of pavement. It was first developed by US Army Corps of Engineers and was verified and adopted by Federal aviation Administration (FAA) and USA Naval Facilities Engineering Command. This index was first developed to evaluate the pavement condition of airports and currently it is used for condition assessment of asphalt-surfaced pavements and plain and reinforced jointed Portland cement concrete pavement. Although it does not provide measurement of skid friction and roughness of the surface and the structural capacity, the PCI provides an objective and rational basis for repair needs and priorities. The PCI is based on visual inspection of a sample unit which is belonging to a section of a branch of the entire network. The PCI rating system is ranged between 0 and 100 where 0 to 10 is classified as Failed and 86 to 100 is classified as excellent.

Equipment used in this test is very simple and include hand odometer wheel, straightedge or string line with scale and datasheet. The minimum number of sample unit that inspected within a section should be statistically calculated according equations. Details of calculations can be found in ASTM D5340. However, some agencies used the following table to determine the number of sample units:

1) 1 to 5 sample units for 1 sample unit

2) 6 to 10 sample units for 2 sample units

3) 11 to 15 sample units for 3 sample units

4) 16 to 40 sample units for 4 sample units

5) Over 40 sample units used 10 %

Although the procedure of calculations of PCI is somehow omplicated and is different for AC and PCC, the PCI provides xcellent feedback on the performance of the pavement for nprovement or validation of maintenance procedures.

5.12.4 Pavement Serviceability Rating (PSR)

The pavement serviceability rating is defined as the judgment of the road's users on the present serviceability of the road at a point in time. This kind of evaluation is based on the concept that the performance of the road at a point in time is what level of service the road provides to users.

AASHTO suggests that the users rate the present serviceability from 1 to 5 and as follows:

4.0 – 5.0: Only new (or nearly new) superior pavements are likely to be smooth enough and distress free (sufficiently free of cracks and patches) to qualify for this category. Most pavements constructed or resurfaced during the data year would normally be rated in this category.

4.0: Pavements in this category, although not quite as smooth as those described above, give a first class ride and exhibit few, if any, visible signs of surface deterioration. Flexible pavements may be beginning to show evidence of rutting and fine random cracks. Rigid pavements may be beginning to show evidence of slight surface deterioration, such as minor cracks and spalling

3.0: The riding qualities of pavements in this category are noticeably inferior to those of new pavements, and may be barely tolerable for high-speed traffic. Surface defects of flexible pavements may include rutting, map cracking, and extensive patching. Rigid pavements in this group may have a few joint failures, faulting and/or cracking, and some pumping.

2.0: Pavements in this category have deteriorated to such an extent that they affect the speed of free-flow traffic. Flexible pavement may have large potholes and deep cracks. Distress includes raveling, cracking, rutting and occurs over 50 percent of the surface. Rigid pavement distress includes joint spalling, patching, cracking, scaling, and may include pumping and faulting.

1.0: Pavements in this category are in an extremely deteriorated condition. The facility is passable only at reduced speeds, and with considerable ride discomfort. Large potholes and deep cracks exist. Distress occurs over 75 percent or more of the surface.

Results of PSR evaluation should be correlated. This may include only variables that can be sensed and measures including cracking, spalling, and others.

5.12.5 Present Serviceability Index (PSI)

PSI can be defined as algebraic function of PSR. In other word: it is a multiple regression analysis to find correlates between PSR an PSI. The relationship between PSI and PSR can be written in th form of

$$Y = A0 + A1X1 + A2X2 + \ldots\ldots\ldots. + E \quad \ldots\ldots\ldots\ldots\ldots\ldots\ldots\ldots\ldots(3)$$

Where:

Y is an average PSR for a pavement section

X1, X2… are correlates between PSR and PSI

E is the discrepancy between PSR and PSI

So the multiple regressions aim at reducing the value of E. Other researchers suggest calculating the PSI relying on roughness, rutting depth and major cracking and as follows:

For rigid pavement

$$PSI = 5.41-1.8 \log(1+SV)-0.09 (C+P)1/2 \quad \text{……………….(4)}$$

For Flexible pavement

$$PSI = 5.03 - 1.8 \log (1+SV) - 0.01 (C+P)1/2 \; RD2 \quad \text{……….(5)}$$

Where:

C is the degree of cracking measured by m2/1000m2

P is degree of patching measured by m2/1000m2

RD is rutting depth by cm

SV is numerical regressions for a profile

It should be remembered that PSI is not a measure of pavement performance by itself. The repeated reading of PSI in different times will provide projection of time –serviceability which can be used for comparison with the required or accepted performance. The AASHTO suggests a biweekly measurement program. Accordingly, the PSI can be computed biweekly and time-serviceability projection can be monitored and used to design the future maintenance and cost-effective programs.

5.12.6 Maintenance Condition Index (MCI)

This index is very similar to PSI and it was first developed in Japan. The MCI is mainly depending on the evaluation of highway engineers, road users and pedestrian to three main parameters of the pavement naming longitudinal roughness, cracking and rutting. This index was introduced as a condition assessment evaluation on one hand and a tool of decision-making on the other hand. The correlations of the three factors can be calculated according to the following equation:

$$MCI = 10 - 1.48\ CR0.3 - 0.29\ RD0.7 - 0.47\ LR0.2 \quad \ldots\ldots\ldots\ldots(6)$$

Where:

CR is cracking as a percentage

RD is rutting depth by mm

LR is the longitudinal roughness by mm

Results of the equation can be compared to the following rates

1) MCI greater than 5 means that the pavement in excellent conditions and needs no maintenance

2) MCI between 4 and 5 means the pavement in good condition and plans of rehabilitation should be started

3) MCI between 3 and 4 means the pavement condition is acceptable and minor rehabilitation is required

4) MCI lower than 3 means the maintenance should be started

5.12.7 International Roughness Index (IRI)

The IRI is the most commonly worldwide used index in evaluation of pavement condition. The index was introduced in 1986 based on works of United States National Cooperative Highway Research Program (NCHRP) at early eighties. The index is currently used by World Bank in managing pavement and also in evaluating new construction and in determining penalties or bonus payments for contractors.

The IRI is defined as the measurement of variation in elevation along a slice of pavement. It can be calculated using any valid measurement tools that range from static road survey to high speed inertial system. Currently, a Quarter-car vehicle math model as shown in the figure below is used to measure the accumulated roughness of road by m/km or in/mi. The main advantages of this method are being stable with time, repeatable and reproducible. The IRI is calculated by simulating of the response of "golden car" having a speed of 80 km/hr,

This method is covered by ASTM E1926-08 and E1364-95 and others and it currently used in most countries around the world to assess the goodness of the new and old pavement. Evaluation of results of this method is performed using the figure of scale shown below.

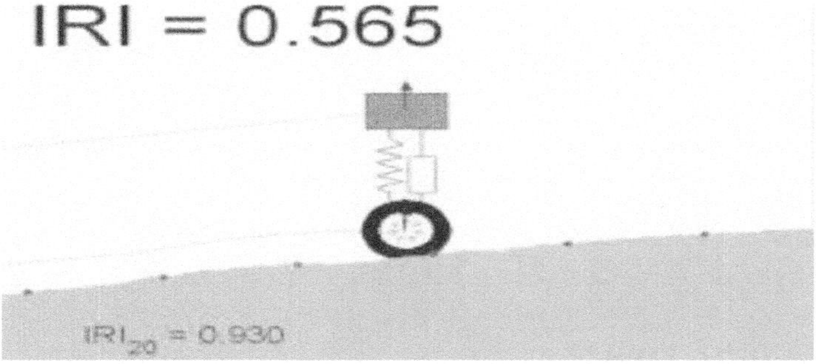

Figure (5.11): Quarter – Car Model used in IRI

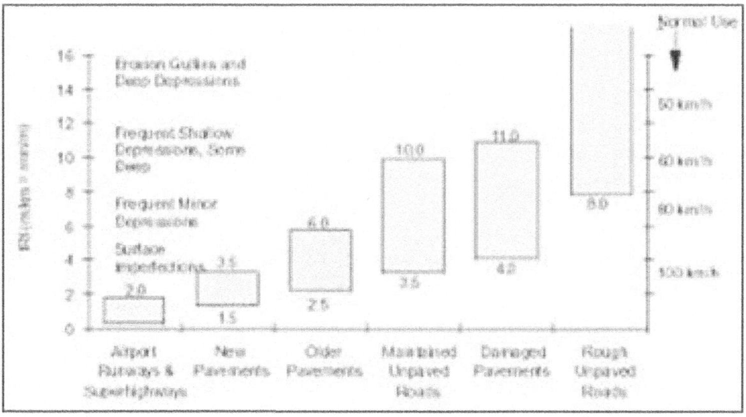

Figure (5.12): IRI Scale

5.12.8 Pavement Quality Index (PQI)

One of the most adopted index to plan for maintenance in North America. The index is ranged from 0 (bad condition) to 100 which is the excellent condition. The data used to calculate PQI is the smoothness of roadways and any signs of distress. The measure is normally repeated each three years. Results of the PQI are also used in setting the priorities of maintenance and also in selecting the best maintenance method such as using filler materials or replacing the entire surface or other as explained in chapter five of this book.

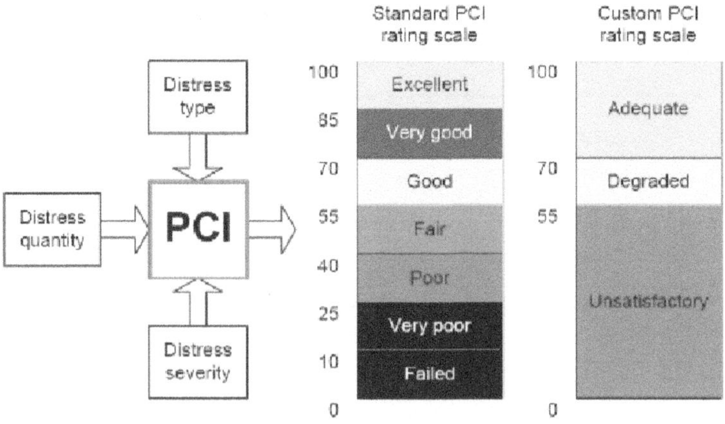

Figure (5.13): Pavement Quality Index (PQI)

5.13 Methods of Pavement Condition Survey

Pavement quality surveys can be defined as the collection of data of serviceability and structural integrity of pavement. This will be used in computing the condition assessment indices whether estimated or measured indices. The survey of pavement condition consumes time and resources and it can play the vital role in the calculation of pavement condition. In addition to its role in making the most efficient maintenance program, the survey of pavement condition is also greatly help in prioritizing and comparing between the different roads.

In general, there are two methods in collecting data of pavement condition: manual and automated data collection. It should be stated that most agencies are moving towards the automated data collection.

A comparison between the two approaches are listed in the following table (5.4).

As it can be seen in the table and except for the cost, the automated collection survey had many advantages and can be accurately used to provide the right condition assessment of the pavement.

Table (5.4) Comparison Between the Two Approaches

	Automated Data Collection	Manual Data Collection
Time	lower data collection times	Longer data collection times
Safety	safer means of collecting data	Personnel at risk collecting data
Objectivity	Objective measurements	Usually subjective since it depends on experience of personnel
Data Size	Vast amounts of data collected & stored depending on capacity of equipment	Agencies may only be able to collect smaller amounts of data at a time

Data Handling	Not subject to transcription errors	Subject to transcription errors
Employers	Suitable in agencies seeking to downsize number of employees	Source of employment for rating staff
Coverage	May cover footprint of data collection vehicle. Multiple runs sometimes needed to cover entire road width	Inspectors can cover entire width of road section relatively easier
Cost	Very expensive equipment costs	Relatively less expensive

5.14 Pavement Maintenance Management System

Highways represent a huge investment; thus it is very important to protect them. A Pavement Maintenance Management System (PMMS) can be the basis tool to help in protecting these highways. This tool can help decision-makers to preserve these highway assets. PMMS can be defined as a systematic methodology to assist in the process of decision making related to evaluate and maintain pavements in an acceptable level. Its goal is to simplify the coordination of activities and evaluate the consequences of decisions in an effective manner in order to maintain pavements in the best condition with available funds.

The PMMS is a tool that can be used to make informed decisions about the maintenance and rehabilitation of a pavement network. The American Public Works Association (APWA) describes the PMMS as "a systematic process for routinely collecting, storing, and recovering the kind of decision-making data required to make use of limited maintenance (and construction) dollars" The American Association of Highway

Transportation Officials (AASHTO) states that the "function of a PMMS is to improve the efficiency of decision making, expand its scope, provide feedback on the consequences of decisions, facilitate the coordination of activities within the agency, and ensure the consistency of decisions made at different management levels within the same organization" .

Schacke and Ertman (1987) mentioned that there is no universally accepted definition of PMMS and some countries include in their definition of a PMMS pavement design as well as construction, maintenance and rehabilitation, while other countries only include construction, maintenance and rehabilitation which in some countries is named PMMS (Pavement Maintenance Management System).

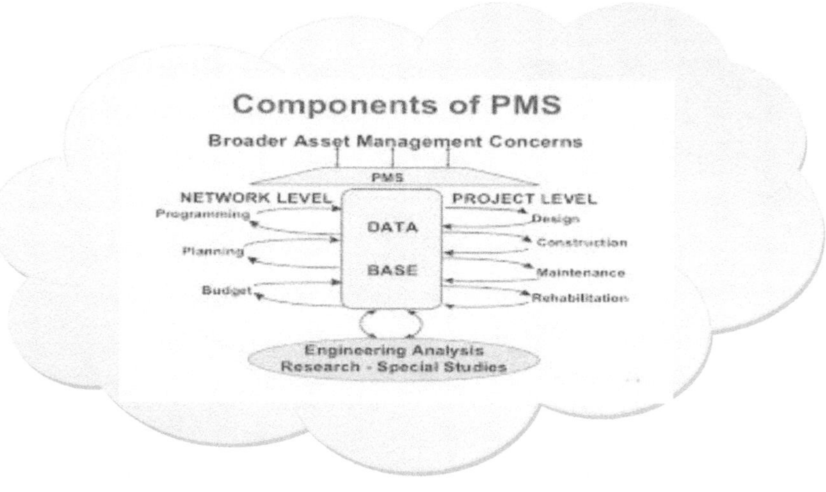

5.14.1 Benefits of PMMS

There is a set of benefits in using the PMMS instead of using judgment and engineering experience alone. Al-Hallaq (2004) and Margaret (1996) summarized these benefits as follows:

It facilitates the process of decision-making and increases the chance of making decision in an optimal way.

It provides an inventory that includes information on location, type of pavement, pavement area, functional classification, mileage, etc.

It shows the present condition of the pavement network depending on systematic and sound engineering processes for obtaining objective pavement condition data.

It helps to forecast the network condition up-to-date, as a function of the funds allocated to make improvements.

It improves the effectiveness of dollars spent on the pavement network.

It defines an estimated budget needed to transport the total roadway network from its existing condition to desired

It defines estimated budgets to preserve a roadway network at specific limits of performance for many years (5 to 20 or more) based on the level of sophistication integrated in the system.

It lists methods to prioritize expenses when funding is less than needed to meet specific performance goals.

It serves as a base for making comparison between alternate preservation strategies for pavement maintenance, rehabilitation, and reconstruction within the network.

It provides a means to monitor the condition of pavement network and provides a quantifiable rating of network condition.

It provides a comprehensive database which contains data relating to pavement condition, construction, maintenance and rehabilitation histories, traffic levels, and any additional data that may be required or specified.

5.14.2 Components of PMMS

PMMS consists of two major components :

1) An information system to manage, collect, and store data.
2) Decision Support System (DSS) for processing and analyzing these data for decision making.

The use of the terms like PMMS and PMS can cause a confusion because systems which are produced by different companies can often have quite different properties, even though the same type is referred by these systems. The following definitions will be used here:

1) Information system includes network information;
2) Decision support system (DSS) includes:
    a) Planning system;
    b) Programming system;
    c) Preparation system; and
    d) Operations system.

The general objectives of these terminologies can be described as follows:

a. Network information

An information system, as defined above, that includes information about the highway network.

b. Planning system

It is a decision-support system (DSS) for strategic planning undertaken to produce long term plans for all the highway network; five years or more planning time horizons are undertaken to determine the implications resulting from meeting goals in terms of consequential pavement conditions, user costs, future budget needs.

c. Programming system

Decision-support system (DSS) for programming or tactical planning is concerned with identifying the needs in the budget year; one to three years planning time horizon including identification of sections from the network which require maintenance and the timing of maintenance.

d. Preparation system

Decision-support system (DSS) for project preparation consists of project formation and design, cost estimation, work orders or the preparation of contract and issue.

e. Operations system

Decision-support system (DSS) for the operations management on a daily or weekly basis includes assigning work to be carried out, developing sufficient costs for this work in terms of materials, labor, and equipment, and making arrangements for carrying out it by contract, recording work achievement, and using this information for control and monitoring.

The above four types of DSS can also be defined in terms of planning and management operations. When these move from *planning* to *operations*, the following changes in the process normally occur:

1) The sections of highway network considered change from all those included in the network to only those sections where works are to be carried out;
2) The changing in time horizon is being considered from multi-year time horizon, to budget year, and then to the present day or week;
3) The data used are changed from being summary to detailed with full coverage on the highway part covered by the project; and
4) The computerized processes change from automatic to those undertaken by the user who works interactively with the computer.

5.14.3 Fundamentals of PMMS

Fundamental requirements of PMMS include those illustrated in Figure (5-13):

1) Inventory data: These data are essential for PMMS. It is impossible to keep an asset without fully understanding the 'where, what, when, etc.' of all of it.

2) Condition data: It is essential to know the condition of the asset and the changes in condition over time since the deterioration rates can be quite variable.

3) Maintenance standards: The implementation of maintenance activities should be specified against consistent and appropriate investigatory standards.

4) Economic analysis: Decision-making related to the alternatives of maintenance activities should consider various options, implementation timing, performance modeling, and pay particular attention to the costs of traffic delay.

5) Budgetary analysis: The known availability of funds may invariably affect treatment priorities.

6) Programming: Good programming is essential to ensure that maintenance activities are implemented at the correct time, and avoid unnecessary disruption.

7) Works control and feedback: It is the final step required to finish the management process and to ensure up-to-date records for the system.

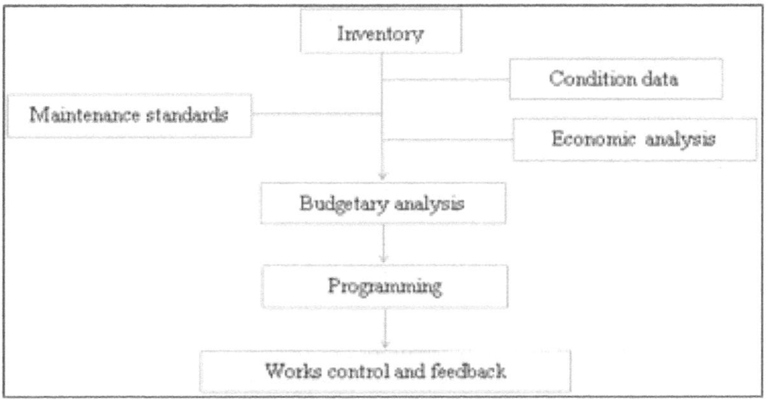

Figure (5-14): Maintenance Management Planning Stages.

5.15 Pavement Maintenance Strategies

All pavement deteriorates over time. Effectively managing the highway network requires optimizing new construction, preservation, maintenance, rehabilitation, and reconstruction pavement needs statewide with multiple constraints, including finite budget requirements, worker safety, user impacts, and regional issues.

Preservation is the preferable network-level pavement management strategy: a long-term, proactive maintenance program intended to maintain pavements in good condition by applying cost effective project strategies at the optimal time to enhance pavement performance, extend pavement life, and conserve resources.

Pavement Management Strategic Plan helps make the best maintenance and renewal decisions and priorities available funding.

The way use the highway network has changed significantly since most of it was built in the late 1940s and 50s, when the largest vehicle expected on the network was approximately 40 tonnes and we had far fewer vehicles on our highway network. In addition, a shift to a service based economy has resulted in an increase of trucks and other commercial vehicles on our highways.

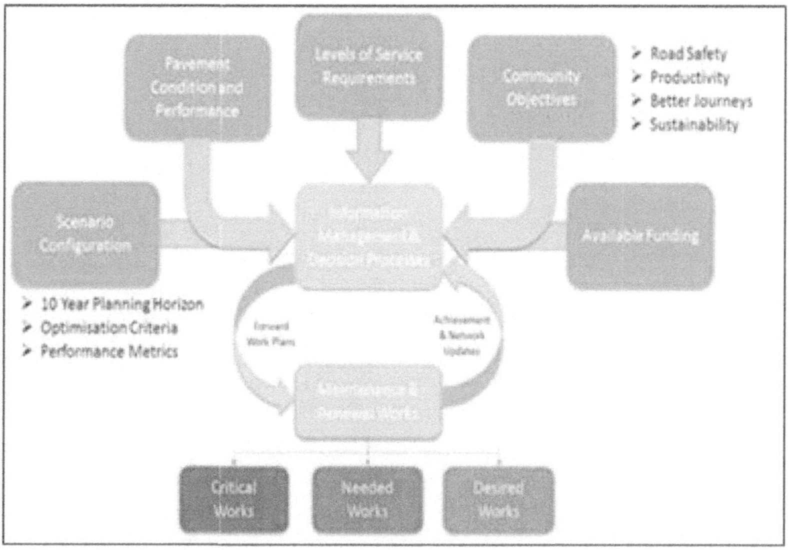

Figure (5.15): Investment Decision Criteria for Maintenance

The Pavement Management Strategic Plan aimed invest in the most important highways at the right time, and ensure looking ahead to what the community will need in the foreseeable future. Pavement Management Strategic Plan has expanded the focus from planning and measuring the work we do, to focus more on community outcomes. Specifically this means investment in highways that contribute to the economy; safety, wellbeing and journey experience.

The strategic plan outlines our investment decision making criteria that are summarized in Figure (5.15)

To be effective, a pavement preservation program must employ the right project-level engineering strategies before the onset of severe distress over an extensive area.

1. Aggregate Surface Treatment (AC Pavement)

Aggregate surface treatment or sometimes called chip seal is consisted of a layer of asphalt binder and a layer of aggregate over the binder layer. The aggregate layer is rolled into the binder. Sand with special grading is sometimes used instead of aggregate, and the treatment technique in this case called sand seal. This method is used to provide a surface seal or skid-resistant surface to structurally sound pavement. This treatment is best suited to low-volume highways.

2. Fog Seal (AC Pavement)

The fog seal technique is normally used in asphalt concrete pavement when the traffic is very light. The fog seal is simply spraying of light bituminous materials on the asphalt concrete pavement using a distributor to improve the waterproofing capability of the pavement and to reduce raveling of the pavement. The density of the bituminous coat material ranges from 0.15 to 0.25 liters per square meters.

3. Rejuvenators (AC Pavement)

Similar to fog seal technique, the rejuvenators is also consisted of spraying of proprietary bituminous material on the surface of asphalt concrete pavement. The main difference between rejuvenators and seal fog techniques is that the betumenous material used in rejuvenators technique penetrates the asphalt surface and soften the binder which consequently reduce the cracks created by temperature change as a result of significant decrease of the hardening rate of the surface of the asphalt concrete pavement.

4. Slurry Seal/Micro Surfacing (AC Pavement)

Slurry seals are normally used to fill cracks that are less than 5 mm wide. Cracks that are wider than 5 mm should be treated prior to applying the slurry seals. The slurry is liquid mixture containing asphalt emulsion, mineral filler, well graded fine aggregate and others special materials. Micro surfacing is similar to slurry seals except that the polymer modified emulsion is used instead of asphalt emulsion. This kind of treatment provides food skid resistance to asphalt concrete pavement. Both slurry seals and micro sealing can be used at the same time and the later is used to provide transverse surface leveling before overlying.

5. Crack Sealing (AC or PCC Pavement)

Crack sealing or resealing is widely used as a preventive and cost-effective measure in both AC and PCC pavements. This treatment should be used when the condition of the pavement is good, otherwise it is not cost-effective in badly deteriorated pavements. This treatment is used in cracks wider that 5 mm and it includes cleaning the cracks thoroughly and sealing them. The technique can be used for both longitudinal and transverse cracks and it can be also used for reflective cracks. The main purpose of the treatment is to hinder the infiltration of surface water into the foundation of the pavement.

6. Full-Depth Patching (AC Pavement)

Full depth patching is a technique that is very suitable to repair distress such as corrugation, rutting and alligator cracking where failure of materials can be extended as far as the sub-base layer of the pavement. The technique can also be used to repair the slippage cracking where the failure of materials is limited to the top surface. In both cases the failed materials are removed and full depth patching is used.

7. Full-Depth Patching (PCC Pavement)

'he full depth patching is also used to repair distresses occurred in PCC slab. This may include the concrete slab or the underlying layers. All the damaged materials related to the distresses is partially or completely replaced. This technique covers a wide variety of distresses such as D cracking and corner breaks. It is also called "slab replacement" when only the concrete slab is replaced. Distresses occurred near joints can also be repaired using this technique considering that the load transfer across the joint to be restored. As previously mentioned, the reflective joint racks of the asphalt concrete can be repaired using the full depth patching.

8. Joint Sealing (PCC Pavement)

Joints occurred in PCC pavement should be thoroughly cleaned and sealed or resealed to prevent infiltration of surface water into the foundation of the pavement. This process is called joint sealing and it is very useful to stop cracks accumulation in the joints.

9. Partial-Depth Patching (PCC Pavement)

Some distresses such as corner spalling is limited to the top of the PCC slab. The partial depth patching, which include removing the damaged materials and replaced with special mixtures, can be used to these local surface distresses. Mixtures that are used to replace the damaged surface materials may include epoxy concrete and special mixtures that provide good skid resistance.

10. Cold Milling (AC or PCC Pavement)

Cold milling is the removal of a given thickness of the surface layer using a machine containing a rotary drum with teeth. It is used in asphalt pavement to bring the pavement grade to an acceptable level, remove a deteriorated layer, and to provide good bonding with the overlay. The removed materials from the repaired pavement can be reuse through blending them with new asphalt and aggregate and reapply the new mixture as a surface layer. The same technique is also used in the rehabilitation of concrete pavement where top surface is cold milled and an overly is applied on a grout coat.

11. Cold Recycling (AC Pavement)

Cold recycling is and environmental aspect that is currently used in sustainable roads. Heating that is used in producing asphalt represent an environmental challenge because of the CO_2 emissions and the huge consumption of fuel. This technique involves using of the waste materials came from the removed pavement as a base course of new pavements. It can also be used as a surface layer in roads of low-volume traffic.

12. Hot Recycling (AC Pavement)

Many outstanding researches proved the feasibility of using waste materials of removed pavement or materials from cold milling as part of conventional hot mix.The waste materials can be used as a replacement of new aggregate which contribute in decreasing the cost of construction and serve the reservation of the natural resources. This kind of pavement (hot recycling) is widely used around the world.

13. Hot In-Place Surface Recycling

A new in-place technique involves mixing of the reclaimed materials with admixtures and reapplying and compacting the new mixture after heating and softening the pavement surface. This can be the top surface of the pavement or can be followed by and overlay. This technique is very useful to rejuvenate the surface and delay the appearance of reflective cracking.

14. Cracking and Seating (PCC Pavement)

This technique is used in deteriorated PCC pavement and it involves cracking the pavement into small pieces. Pneumatic rollers is used to seat the small pieces followed by overlying with asphalt concrete surface. This technique was found to decrease the severity of reflecting cracks.

15. AC Overlay (AC or PCC Pavement)

Required quality or functional requirements of highways can be changed based on changing of some criteria such as volume of traffic. The AC overlay is very efficient cost-effective technique to improve the quality and correct the functional requirements of highways especially when they are in good conditions. It can be applied on both types of highways AC and PCC pavements and in both cases, one or more layers to be added on the surface of the pavement. This technique can be used in combination with other abovementioned techniques such as hot recycling and cold milling.

16. PCC Pavement Overlay (AC or PCC Pavement)

Similar to AC overlay technique, the PCC overlay is also used to improve and correct the functional requirements of the existing PCC pavement. Although the technique is primarily used with PCC pavement, many cases were recorded for the use of PCC overlay on AC pavement. In this technique, one or more layers are added on the PCC or AC pavements.

17. Reconstruction (AC or PCC Pavement)

Economic studies and analyses should be carried out prior to selecting the maintenance method. The complete or partial reconstruction in many cases is considered when the other maintenance methods are not cost effective or when the functional requirements of the highways are significantly changed. Reconstruction method involves removing of some or all layer and construct new different layers appropriate to the new requirements.

18. Seal Coat

Seal coat is used to waterproof the surface, seal small cracks, reduce oxidation of the pavement surface, and improve friction.

19. Double Chip Seal

This method includes an application of two single seal coats. The second coat is placed immediately after the first. This treatment waterproofs the surface; seals small cracks, reduces oxidation of the pavement surface, and improves friction.

BIBLIOGRAPHY

Dissertation and Thesis

Abbas M. B., "**Optimization of Pavement Maintenance Decision Using Automated Pavement Monitoring System**", Ph.D. Thesis, Baghdad University, Iraq, 2015.

Al-Hallaq M. A. Fatah, "**Development of a Pavement Maintenance Management System for Gaza City**" A Thesis Submitted in Partial Fulfillment of the Requirements for Degree of Master of Science in Infrastructure Engineering, Palestine, 2004.

Faiq, M.S., "**Modeling of Transportation Management for Asphalt Concrete Pavement Using Computer Simulation Technique**", M.Sc. Thesis, Baghdad University, Iraq, 2000.

Faiq, M.S., "**The Use of Artificial Neural Network for Estimation Total Cost of Highway Construction Project**", Ph.D. Thesis, Baghdad University, Iraq, 2009.

Ismail, E.A., "**Highway Intersections with Alternative Priority Rules**", Ph.D. Thesis, University of Bradford, England, 1989.

K.A. Al-Saffar," **Study of the Maintenance and Renovation of the Main Building at Umist**", University of Manchester, U.K., 1984.

Nuha H. A., "**A Proposed Maintenance Management System of Iraqi Hospitals**", M.Sc. Thesis, Baghdad University, Iraq, 2015.

Reem, A. M., "**Cost Deviation Management in Highway Projects**", M.Sc. Thesis, University of Technology, Iraq, 2017.

Book and Handbook

AASHTO (1985) "**Guidelines on Pavement Management**" published by American Association of State Highway and Transportation officials, 444 North Capital Street, N.W. Suit 225. Washington, D.C. 20001.

Ann Johnson, P.E. (2000) "**Best Practices Handbook on Asphalt Pavement Maintenance**" Minnesota Technology Transfer (T2) Center / LTAP Program, Center for Transportation Studies, University of Minnesota, 511 Washington Avenue S.E., Minneapolis, MN 55455-0375.

Barrie Chanter and Peter Swallow, (2007) "**Building Maintenance Management**", Second Edition, Britain.

CASSOA (2010) "**Aerodrome Pavement Maintenance Procedures**" advisory circular, CAA-AC-AGA020.

Edwin Neville White, (1975) "**Introduction to Maintenance Planning in Manufacturing Establishments**, United Nations.

Encon Renew, (2011) "**Introduction to Building Maintenance**" brochure (Policy for the maintenance of Queensland Government Buildings, Second Edition.

Faiq M. Al-Zwainy and Ibraheem A. Aidan, "**Flexible Pavement Maintenance Operations**", Lulu Press, Inc. USA. 2018.

Garber N. J and Hoel L. A. (2009) "**Traffic and Highway Engineering**" fourth edition, United State copyright, Cengage Learning, Canada.

Haas, R. and Hudson, W.R., (1978) "**Pavement Management System.**", McGraw-Hill Book Company.

Higgins & Morrow, (1977) "**Maintenance Engineering Hand Book**", 3rd Ed., Hill Book company, U.S.A.

Newbrough, E. T, (1967) "**Effective Maintenance Management: Organization, Motivation, and Control in Industrial Maintenance**", McGraw-Hill, U.S.A.

O'Flaherty, A.M. (2007) "**Highways: The Location, Design, Construction and Maintenance of Highway Pavements**" Butterworth-Heinemann publications, Linacre House, 4th edition, Oxford OX2 8DP, UK.

Oglesby, C.H. and Hicks, R,G., (1982) "**Highway Engineering.**", 4th Edition, John Wiley & Sons. Inc..

Pierre V., (1995) "**Strategy For Setting up a Maintenance Organization in a Developing Country**", Delhi.

TRB, (1975) "**Traffic Flow Theory: A Monograph**", Transportation research board, Special Report 165.

U.N. Economic Commission for Africa, "**Road Maintenance Handbook.**", Vol. III, (Maintenance for paved roads).

Articles

Acra, I., Fil, E., Fim, U.N.E., Fih, E., and Mrtp, I., "**Maintenance Standard for Roads in Rural Country.**", The Highway Engineer, The Journal of the Institution of Highway Engineers, Vol. 31, No. 3, March, p.p. 20-25, 1974.

Jendia S. and Al-Hallaq M. "**Development of a Pavement Maintenance Management System (PMMS) for Gaza City**" Journal of the Islamic University of Gaza (series of natural studies & engineering) Vol.13, No.1, P.119-138, 2005.

Robinson, R., "**Road Maintenance and Management for Developing Countries.**", The Journal of the institution of highway and transportation, Vol. 33, No. 6, June, p.p. 8-12, 1986

Conferences and Congress

C.W. Gits, **Development in the Fundamentals of Maintenance Management**, (International Maintenance seminar - F.L.Smith, 1990.

Hicks R. Gary, Seeds S. B., and Peshkin D. G. " **Selecting a Preventive Maintenance Treatment for Flexible Pavements**" a report prepared for Foundation for Pavement Preservation, Washington, 2000.

Jan H. Nielsen, **Illustration of a Computerized Maintenance System Which May Be Designed to Operate at Various Functional Levels**, (International Maintenance seminar - F.L.Smith, 1990.

Margaret Broten P.E. "**Local Agency Pavement Management Application Guide**" Washington State Department of Transportation, Applied Pavement Technology, Inc., Published by: The Northwest Technology Transfer Center, 1996.

Schacke I. and Ertman Larsen H.J. "**Pavement Maintenance Management Systems in OECD Countries**" proceeding by 2nd North American Management Conference, 1987.

Shahine, M.Y. and Walther, J.A., "**Pavement Maintenance for Roads and Streets Using the PAVER System.**", U.S. Army Construction Engineering Research Laboratory (USACERL), Technical Report m-90/05, USA, July, 1990.

The World Bank "**Selecting Road Management Systems**" World Bank, European Union TACIS Program and GTZ. Proceedings of the Highway Policy Seminar for Countries of the Former Soviet Union, Washington DC: International Bank for Reconstruction and Development/World Bank, 207- 216, 1995.

SIMILAR BOOKS

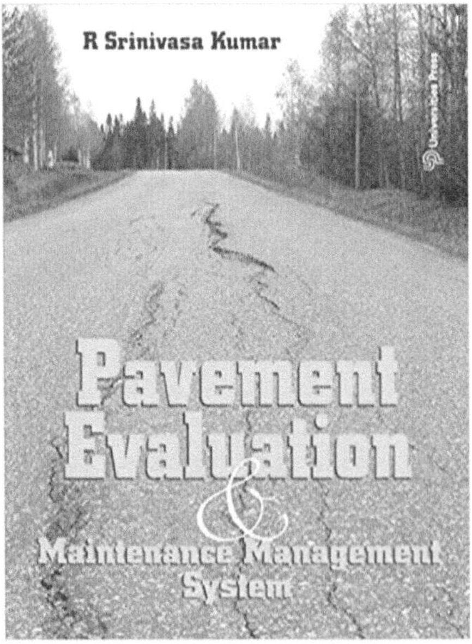

PAVEMENT MAINTENANCE AND MANAGEMENT

Made in United States
Cleveland, OH
06 November 2026

47780404R00143